Rheological Measurement Techniques and Analysis Methods

Edited by Jian Wang

Published in London, United Kingdom

Rheological Measurement Techniques and Analysis Methods
http://dx.doi.org/10.5772/intechopen.1001622
Edited by Jian Wang

Contributors
Camilo A. Franco, Diana A. Estenoz, Esteban Taborda, Farid B. Cortés, Iskandar Bin Dzulkarnain, Jian Wang, Lady J. Giraldo, Muhammad Mohsin Yousufi, Mustapha Bouallala, Mysara Eissa Mohyaldinn Elhaj, Noor Jahan, Rahul Sharma, Yucan Liang, Yue Cheng, Yurany Villada

First published in London, United Kingdom, 2024 by IntechOpen
IntechOpen is the global imprint of INTECHOPEN LIMITED, registered in England and Wales, registration number: 11086078, 167-169 Great Portland Street, London, W1W 5PF, United Kingdom

For EU product safety concerns: IN TECH d.o.o., Prolaz Marije Krucifikse Kozulić 3, 51000 Rijeka, Croatia, info@intechopen.com or visit our website at intechopen.com.

British Library Cataloguing-in-Publication Data
A catalogue record for this book is available from the British Library

Rheological Measurement Techniques and Analysis Methods
Edited by Jian Wang
p. cm.
Print ISBN 978-0-85466-761-1
Online ISBN 978-0-85466-760-4
eBook (PDF) ISBN 978-0-85466-762-8

If disposing of this product, please recycle the paper responsibly.

Meet the editor

Jian Wang is a professor at the Beijing University of Chemical Technology, China, specializing in polymer engineering and science. He earned his Ph.D. in Mechanical Engineering in 2010 and has conducted research at the University of Wisconsin–Madison, USA (2009-2010), Beijing Institute of Technology (2010-2018), China, and RWTH Aachen University, Germany (2017-2020). Prof. Wang is a fellow of the China Plastic Processing Industry Association (CPPIA) and the China Plastics Machinery Industry Association (CPMIA). He has published more than 100 peer-reviewed papers and 60 patents. He has led numerous research projects funded by the Natural Science Foundation of China, the National Key Technology Research and Development Program of China, and various university and enterprise initiatives.

Contents

Preface

Rheology, the science of material flow and deformation, has become increasingly significant across various scientific and industrial fields. As material science continues to evolve, the ability to measure, understand, and predict the behavior of complex materials under different conditions becomes essential. This edited book, *Rheological Measurement Techniques and Analysis Methods*, provides a comprehensive exploration of advanced rheological methods, their applications, and the challenges faced by industries that rely on the analysis of flow behaviors.

Chapter 1, "Real-Time Rheological Measurement Techniques in Plastics Industry" delves into the development of real-time rheological techniques, discussing their vital role in monitoring and controlling processes in the plastics sector. This chapter covers the principles, instrumentation, and applications of various rheometers, such as slit, capillary, and ultrasonic, offering insights into real-time solutions for quality control and product development. In Chapter 2, "Exploring the Rheological Properties of Viscoelastic Materials" the focus shifts to understanding the interplay between viscous and elastic characteristics. This chapter bridges theoretical concepts like the Deborah number and storage modulus with real-world applications, offering a comprehensive overview of measurement techniques and practical applications for materials that fall within the viscoelastic spectrum. Moving on to the theoretical aspects of material behavior, Chapter 3, "Modeling, Analysis, and Numerical Solution of a Viscoelastic Contact Problem with Normal Compliance in the Context of Locking Materials" presents a mathematical investigation of viscoelastic materials in contact with rigid foundations. This chapter highlights a novel viscoelastic model, providing readers with a detailed explanation of variational inequalities, finite element schemes, and error estimation, all crucial for those working with materials exhibiting locking behaviors. The focus on industrial applications becomes even more prominent in Chapter 4, "Emulsion Rheology: Applications and Measuring Techniques in Upstream Petroleum Operations" This chapter provides an in-depth analysis of emulsions' role in the oil and gas industry, where rheological analysis of emulsified working fluids is critical for optimizing extraction and refining processes. Key techniques, including rotational and extensional rheometry, are explored, offering valuable insights into non-Newtonian fluid behavior in petroleum production. Chapter 5, "Application of Nanotechnology in the Petroleum Industry: A View from Rheology" introduces the cutting-edge integration of nanotechnology into rheological studies within the energy sector. The chapter covers significant advancements in using nanomaterials to enhance crude oil transport, improve drilling fluid performance, and increase recovery efficiency, all underpinned by rheological analysis.

This book brings together a wealth of knowledge from different industries, providing readers with essential tools to approach rheological challenges from both theoretical and practical perspectives. I hope that this book serves as a valuable resource for researchers, engineers, and industry professionals engaged in the field of material science.

I would like to extend my heartfelt thanks to all contributors whose insights and expertise have greatly enriched this book. Special thanks go to the editorial team, whose efforts have been instrumental in bringing this collection to fruition. I would also like to express my deepest gratitude to my wife and son, whose unwavering support and understanding during the hours of editing have been invaluable. Their patience and encouragement have allowed me to dedicate the time and effort needed to bring this book to completion.

Jian Wang
Beijing University of Chemical Technology,
Beijing, China

Section 1

Rheological Measurement in Plastic Industry

Chapter 1

Real-Time Rheological Measurement Techniques in Plastics Industry

Yucan Liang, Yue Cheng and Jian Wang

Abstract

This chapter aims to provide a comprehensive overview of real-time rheological measurement techniques and their applications in various fields. Rheology, as the study of the flow and deformation of materials, plays a vital role in understanding the mechanical properties and behavior of complex fluids and soft matter. With the advancement of technology, real-time rheological measurements have become increasingly crucial for *in situ* monitoring and control applications. This chapter discusses real-time rheological measurement techniques' principles, instrumentation, and applications.

Keywords: real-time rheology, online rheology, inline rheology, slit rheometer, capillary rheometer, oscillatory rheometer, ultrasonic rheometer

1. Introduction

Polymer products are utilized extensively across diverse applications spanning the automotive, consumer goods, aerospace, electronics, biomedical, and mineral processing sectors. Specific sectors impose stringent quality standards on polymer products. Commonly used methods for forming polymer material products include additive manufacturing, injection molding, extrusion molding, and blown film molding. For additive manufacturing (AM), the polymer material is melted and extruded and then stacked layer by layer to complete the processing of the product [1]. For injection molding, the polymer material is molten, injected into a specific mold, and then cooled to form the product [2]. Similarly, extrusion molding is the continuous process by which a molten polymer material is extruded from a hopper to a head with a fixed runner and then cooled to form a product [1]. A polymer, characterized by a linear molecular chain arrangement, exhibits larger dimensions along the chain than in other directions, setting it apart from other materials with its intricate rheological behaviors [3]. The intricate flow channels and inherent polymer characteristics significantly impact material flow and deformation during processing. Rheological testing helps to understand the viscoelastic behavior of polymer materials. It is used to guide the setting of process parameters, screw design optimization, and computer simulation calculations for the processing of polymer products [4]. In addition,

IntechOpen

rheological measurements are also suitable for evaluating new materials and improving the composition of materials to enable more rapid development [5].

In material extrusion additive manufacturing, the product is formed in three steps. Firstly, the filament material or solid particles are fed into a heating unit. Secondly, the material is heated to a temperature point with a high flow state. Lastly, the material is dispensed onto a printing platform to solidify and form the product, depending on the parameters of the product model [6]. The polymers always exhibit viscoelastic behavior during the material extrusion additive manufacturing process. At the same time, the printability of the material and the quality control aspects of the printed product are highly related to the rheological properties of the polymer material. Inadequate viscosity during extrusion from the nozzle may affect the mechanical and structural characteristics of the product. In addition to this, many semi-crystalline polymers, when extruded from the extrusion head, not only affect the interfacial diffusion between the deposited layers due to rapid crystallization but also lead to a high level of residual stresses inside the printed part, causing shrinkage and warpage of the product. Controlling the fluctuation of melt viscosity during 3D printing can help to solve the above problems. Still, unfortunately, the current parameter tuning method commonly used in additive manufacturing is a trial-and-error approach based on time and cost.

Injection molding is a mature technology that has been in use for decades, and the quality of the product is related to the processing parameters, the processed material, and the plasticizing quality [7]. In particular, plasticization quality depends on the shear rate, temperature, and pressure during the plasticization phase, which ultimately manifests itself in the viscosity of the melted polymer material. During the injection phase, when the material is injected into the cavity, changes in its viscosity can lead to flow behavior inconsistent with that of the last injection, and differences in pressure changes along the flow direction may produce different degrees of shrinkage, which may ultimately lead to warpage of the part. Similarly, for extrusion molding, rheological assessments play a pivotal role in both scientific research and industrial applications [8]. Insufficient time for polymer molecules to relax during processing may give rise to die swell or other instabilities, particularly pronounced in more branched and higher molecular weight polymers requiring extended relaxation times. Modulating melt flow and heat transfer parameters, such as barrel temperature, pressure in the mold cavity, or drive power, provides a means of controlling the injection molding process.

Whether it is additive manufacturing, injection molding, extrusion, or any other method of processing plastic products, a standard method of testing melt viscosity is by using commercial rheometers. This type of measurement is a non-real-time rheological testing method, where samples of material produced from the factory need to be sent to a laboratory for testing, and the data obtained is used to correct process parameters on the production line, with a significant time lag. During this time, it may lead to an increase in cost [3]. Real-time rheological measurements aim to characterize the rheological behavior of materials during processing. However, offline testing methods do not provide direct access to the rheological parameters of the material in the mold or the barrel, and the results it obtains may produce errors depending on the actual processing conditions [9]. However, offline testing has become a standard test, which has the research value of allowing direct comparison of the flow characteristics of different materials.

At present, the era of Industry 4.0 is coming; the digitalization and real-time monitoring of the production process are becoming more and more critical, and the adaptive control of the production process using big data and artificial intelligence will become a hot spot of research in the future [8]. Acquiring real-time rheological parameters will become an essential part of intelligent injection molding or other industrial automation. This chapter mainly introduces the standard methods of real-time rheological measurements and their application areas, laying the foundation for future research in related fields.

2. Principle of real-time rheology

With the development of colloid (now called polymer) chemistry in the first half of the nineteenth century, scientists saw the need for a separate discipline to describe the fluid properties of polymers as becoming increasingly urgent. Hence, the concept of the present-day discipline of rheology [10] and the Society of Rheology were established at the end of 1929. One of the most important symbols of the birth of rheology is the shear-thinning phenomenon. In 1925, Ostwald [11] observed that different shear rates make the materials show different viscosities. In 1929, the thesis of Hencky [10] proved this phenomenon. Classical Newtonian fluid mechanics posits that viscosity remains unaffected by shear rate [11]. Consequently, the emergence of rheology as a distinct branch of physics can be attributed to the recognition of shear-dependent viscosity. The substances responsible for this phenomenon are alternatively referred to as non-Newtonian fluids. The rheological properties of materials are divided into two main aspects: on the one hand, the study of the relationship between viscosity, shear rate, temperature, time, and other influencing factors; on the other hand, the dynamic viscoelasticity [12–14]. Currently, the most commonly used polymer viscosity model is the Carreau equation [15] (Eq. (1)):

$$\eta = \eta_\infty + \frac{(\eta_0 - \eta_\infty)}{\left[1+(\lambda\dot{\gamma})^a\right]^{\frac{(1-n)}{a}}} \tag{1}$$

where η_0 is the Newtonian viscosity, η_∞ is the shear viscosity at a much high shear rate, α is the characteristic time, inverse α is the critical shear rate (starting Newtonian fluid behavior), λ is relaxation time, n is the non-Newtonian index, and $\dot{\gamma}$ is the shear rate.

The most mainstream research method for the effect of time and temperature on viscosity is based on the time-temperature superposition principle (TTSP) [16]. The viscosity curve of materials, subject to rapid variations at low temperatures and short durations, can be extrapolated to encompass high temperatures and prolonged durations. This process facilitates deriving a comprehensive viscosity master curve for the material across an extended testing range. The most widely used temperature conversion factor is based on the Williams-Landel-Ferry equation [17] (Eq. (2)):

$$\log(\alpha_T) = \frac{-C_1\left(T - T_g\right)}{C_2 + \left(T - T_g\right)} \tag{2}$$

where C_1 and C_2 are empirical constants whose product is about 900.

Rheometers used for viscosity profile testing are mainly compressional rheometers (slit rheometers [18–20] and capillary rheometers [21, 22]) and rotational rheometers (plate rheometers [23] and cone-plate rheometers [24]). Compressional rheometers are mainly used to obtain viscosity profiles at high shear rates. In contrast, rotational rheometers are used to obtain viscosity profiles at low shear rates, and the combination of the two can result in a complete viscosity profile of the material.

On the other hand, the dynamic viscoelasticity of polymers is also an essential rheological property. The primary focus of the research revolves around examining the interplay between storage modulus, loss modulus, and tanδ across varying temperature conditions. The equations for the calculation of these parameters and the relationship are as follows [25] (Eqs. (3)–(5)):

$$G' = \frac{\sigma_0}{\gamma_0} \cos\delta \tag{3}$$

$$G'' = \frac{\sigma_0}{\gamma_0} \sin\delta \tag{4}$$

$$\tan\delta = \frac{G''}{G'} \tag{5}$$

where G' is the energy storage modulus, G'' is the dissipation modulus, and δ is the dissipation angle.

Compared with viscosity testing, dynamic viscoelasticity testing is mainly carried out using an oscillatory rheometer, where the linear dynamic viscoelastic master curve of a material is obtained by applying oscillatory shear at different temperatures to understand the viscoelastic properties of the material in different intervals [25].

Rheometers can be classified into four types, offline, at-line, online, and inline, according to their position concerning the production line and how they are installed [9], as shown in **Figure 1**.

Both offline and at-line rheological testing techniques are non-real-time measurement methods. Offline rheological measurement techniques involve transporting raw materials or products from the factory to a laboratory or measurement center for measurement. This method usually takes more time due to the type of measurement

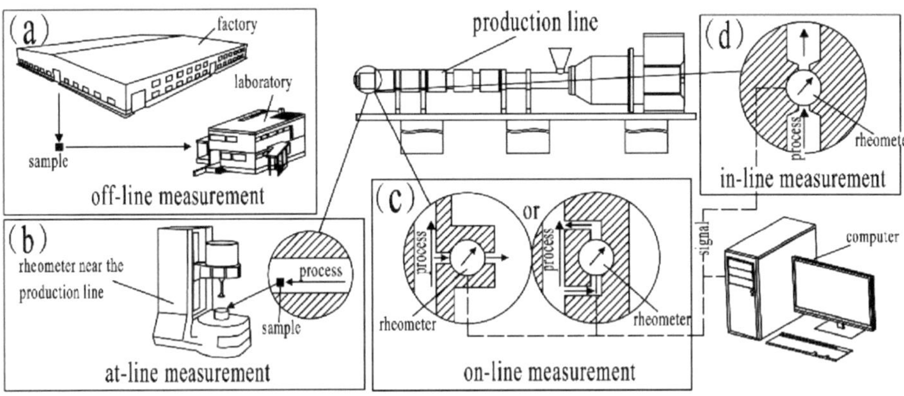

Figure 1.
Four kinds of rheometers: (a) offline rheometers; (b) at-line rheometers; (c) online rheometers; (d) inline rheometers.

and the need to transport the material. At-line rheological measurements use instruments close to the production line so that the operator can remove the material from the production line in time for the measurement to be made in the test instrument. At-line measurements reduce the transport time of the material, but the equipment used in the measurement method is a non-specialized plastometer.

Online rheological testing is generally used on extruders mounted in the extruder's plasticizing unit or extruder head/die [26, 27]. Test instruments using online rheological method are connected to the main melt flow path of the extruder through a bypass. Typically, the molten polymer in the primary flow channel is introduced into a capillary tube through the implementation of a gear pump within the bypass, or the melt in the main flow channel is used directly to feed the material under its pressure into a slit or two rotating disks for measurement. There is a risk that the measured material in the test setup will not be utilized for secondary purposes, thus resulting in unnecessary waste. Some designs [26] allow the material on the bypass to be recycled back into the main flow path, but only online capillary rheometers or online slit tube rheometers equipped with gear pumps can fulfill this function.

Inline rheological testing can be applied on extruders and injection molding machines [9]. In the case of injection molding machines, the device is located inside the nozzle or mold; in the case of extrusion molding, the measuring device is an extruder head with pre-designed channel dimensions [28, 29]. The main principal methods of using inline rheological measurements are capillary rheometers and slit rheometers.

Incorporating real-time rheological measurement devices allows for the assessment of the rheological characteristics of the molten material throughout the processing stage, facilitating the optimization of parameters such as temperature and pressure. The main advantages of real-time rheological measurements over non-real-time measurements are as follows [26]: (1) automated process control, mitigating or eliminating manual intervention and reducing the time lag between sampling and test results, thereby expediting material or process development; (2) sensitivity to minute changes or trends in material properties that may indicate potential product failure; and (3) the systematic collection of pertinent material data at specific intervals to establish correlations among equipment characteristics, operating conditions, material formulations, chemical transformations, and the resulting material's morphology and properties.

Qwist et al. [30] innovatively designed and manufactured a pressure difference apparatus (PDA) capable of producing gels with varying viscosities by conducting online rheological parameter measurements. The PDA demonstrates superior measurement speed compared to offline capillary rheological devices commonly used in this field. In a study by Bonino et al., the cross-linking reaction of UV-capable methacrylate alginate was investigated using a dynamic rheological approach. Real-time *in situ* online rheological testing revealed a power-law relationship between UV intensity and gelation time, exploiting the sudden viscosity drop near the gel point to expedite production process development [31].

Furthermore, offline testing, involving equipment like capillary rheometers for non-real-time measurements, presents challenges due to low piston loading during melt extrusion. This results in pressures that deviate significantly from actual processed pressures and fail to accurately reflect the polymer's melt viscosity during processing. On the other hand, real-time rheological measurements circumvent such issues [9].

3. Instrumentation for real-time rheological measurements

Real-time rheological test instruments mainly include molds/frames that constitute the instrument's main body, temperature/pressure sensors, different shapes of the test area, power modules, and other components (different for different types of rheometers). In this chapter, we mainly classify the instruments and technologies according to the testing principle of the rheometer, which can be divided into the following four types: (1) slit and capillary rheometer; (2) rotational rheometer; (3) oscillatory rheometer; and (4) ultrasonic rheometer.

Slit rheometers and capillary rheometers are both extrusion rheometers. The difference between the two is that the test area of the slit rheometer is a rectangular or other special-shaped flat gap, so it is called a slit rheometer. In contrast, the test area of the capillary rheometer is a cylindrical gap. The main components of the slit rheometer are the slit mold body, sensors, slit test area, etc. The schematic diagram is shown in **Figure 2**. The slit rheometer is affixed either at the exit point of the material processing or at an intermediary bypass location through a connecting flange, featuring a configuration wherein a minimum of 2 or more pressure sensors are seamlessly integrated into the slit test area.

Rotational rheometers are typically categorized into flat and cone-plate types, distinguished primarily by the upper and lower clamping plate configuration. Online rotational rheometer's main components: upper and lower plates, motor, sensor, cleanup ring, etc., the schematic and work procedure diagram shown in **Figures 3** and **4**, the rotational rheometer through the connecting flange is installed in the material processing of any bypass position, from the mainstream to take out enough material for the test, through the baffle plate will be the bypass and the mainstream to isolate the upper and lower plates to move up, the movement of the test position, the cleanup ring shown in **Figure 4(e)** will be the removal of excess material, and at the same time to ensure the sealing of the test area.

The oscillatory rheometer closely resembles the plate-type rotational rheometer, where the lower plate is typically stationary, and the upper plate experiences small amplitude vertical vibrations. By applying an alternating stress field to the material through this setup, one can extract the dynamic viscoelastic characteristics of the material. The main components are also the same as the rotational rheometer, and the difference only lies in that the rotational rheometer motor is rotating motion, and the oscillatory rheometer is a small range of linear reciprocating motions.

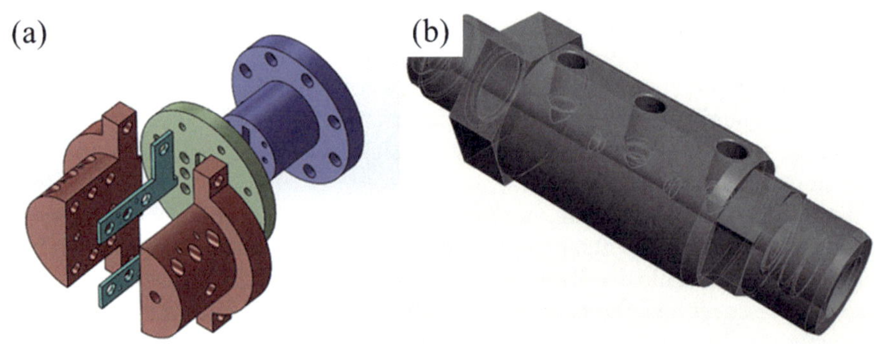

Figure 2.
(a) Real-time slit rheometer [28]; (b) real-time capillary rheometer [32].

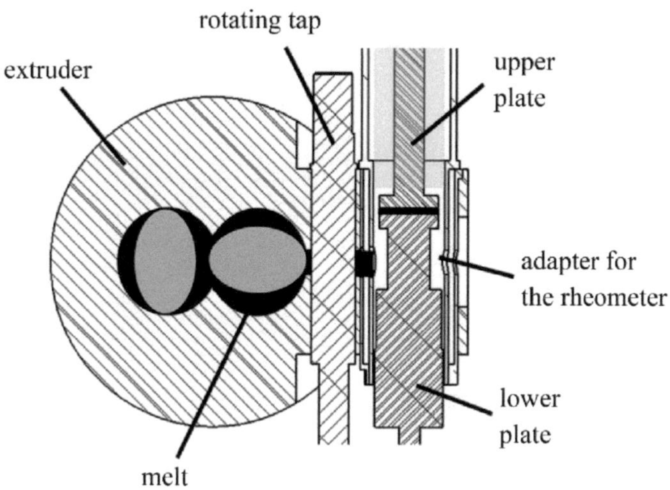

Figure 3.
Online rotational rheometer [4].

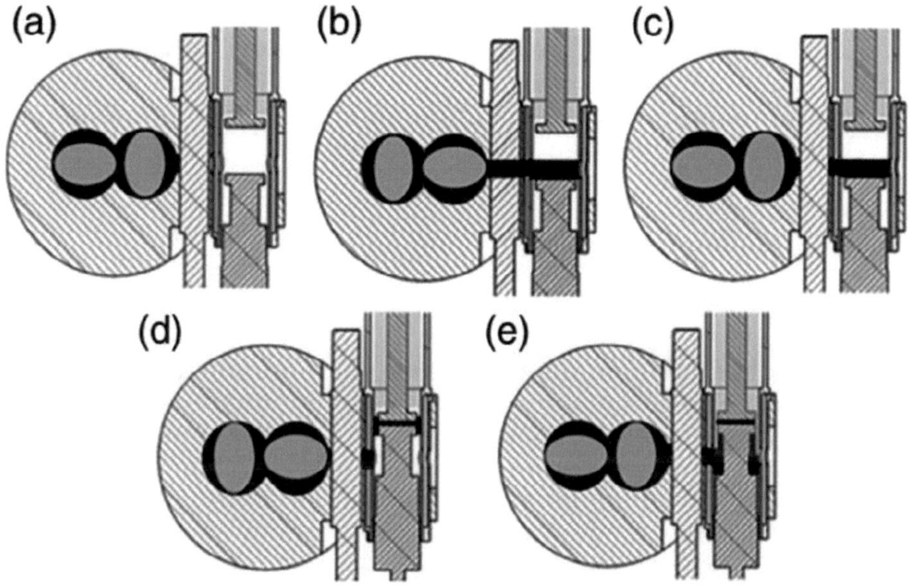

Figure 4.
Real-time rotational rheometer [4].

Inline ultrasonic rheometers are a very new method of non-destructive testing with a very simple composition, usually consisting only of a fixed frame and an ultrasonic transducer, as shown in **Figure 5**, which reflects the flow rate of non-Newtonian fluids in the pipeline using an ultrasonic transducer.

The real-time rheological test instrument's design must adhere to the following requirements: (1) facilitate the extraction of samples representative of the material under processing; (2) ensure that sampling procedures do not interfere with the ongoing processing operations; (3) guarantee that sampling activities do not alter the inherent properties of the material; (4) enable rapid and repeatable execution of

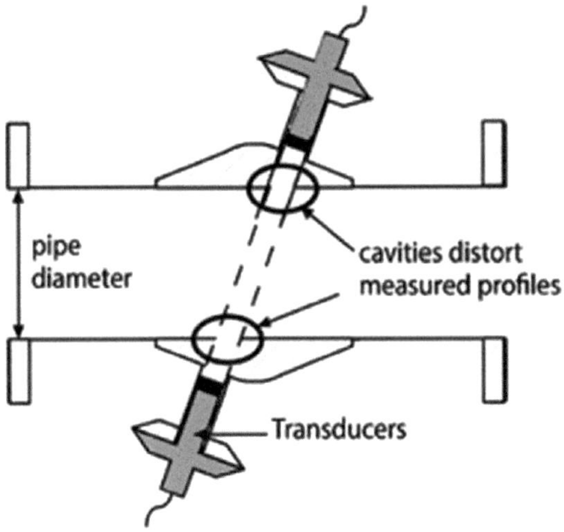

Figure 5.
Inline ultrasonic rheometer [33].

sampling tests; (5) incorporate a sensor with high sensitivity; (6) achieve a high rate of data acquisition; (7) shield the sensor from the adverse effects of processing environments, including high temperatures, high pressure, and mechanical vibrations; (8) demonstrate commendable sealing of the test area; (9) be easily cleanable; and (10) exhibit good ease of use during testing procedures.

4. Online and inline rheometers

4.1 Real-time slit and capillary rheometers

Both slit rheometers and capillary rheometers operate based on the principle of evaluating the rheological characteristics of a material through the measurement of the correlation between pressure drop and flow. This assessment is conducted using distinct pressure transducers to monitor the variations in pressure during the flow process. The test conditions are generally high shear rates, that is, actual processing conditions. The real-time slit rheometer is available in both online and inline configurations. However, its installation position can be at the die head or the main flow path bypass, as shown in **Figure 6**.

In 1991, Pabedinskas [35] designed and built an inline slit rheometer, which at this time was still in its most straightforward stage. The rheometer was mounted on the die head of the extruder with a vertical cone-shaped test area (as shown in **Figure** 7(a) and a rectangular cross section. Three pressure and two temperature sensors were distributed above and below the test area. The outcomes favorably concur with those obtained from a commercially available capillary rheometer. Likewise, a similar online slit rheometer was installed on the injection molding machine for rheological testing of the injected material [36]. The slit rheometer was mounted on the gate for material injection (instead of the original mold), as shown in **Figure** 7(b). As the slit rheometer needs to be installed at the material outlet position, the extruder die/injection mold needs to be replaced by the slit rheometer for rheological testing, which brings no change to the material processing. Teixeira [37] divided the main

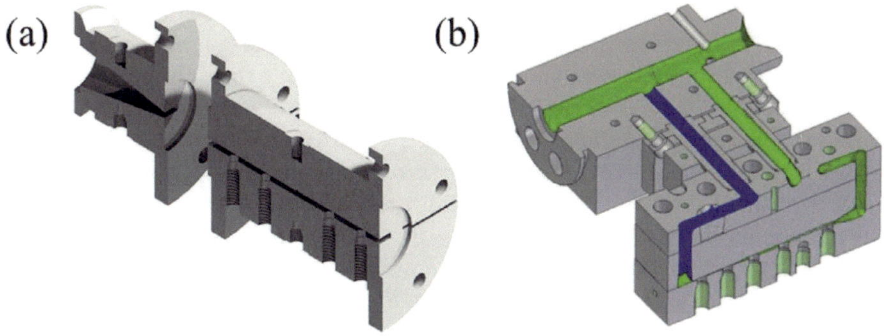

Figure 6.
Real-time slit rheometers: (a) die head slit rheometer (inline form) [34]; (b) bypass slit rheometer (online form) [8].

flow path of the extruder into two identical flow paths (as exhibited in **Figure 7**(c). The slit rheometer was installed in the die head. One flow path was used to test the viscosity of the material, and the other was used to perform the material extrusion process. Nevertheless, this enhancement is contingent to a certain degree on alterations induced in the original material flow characteristics. Luger [8] proposed to be installed in the die position of the slit rheometer converted to the side of the main flow channel and from the main flow channel to lead out of the part of the bypass for rheological testing, and finally through the gear pump will be tested after the bypass of the material flow convergence of the main flow channel.

There are still unavoidable shortcomings of Pabedinskas's, Teixeira's, and Luger's rheological measuring equipment. (1) The material flow through the slit rheometer

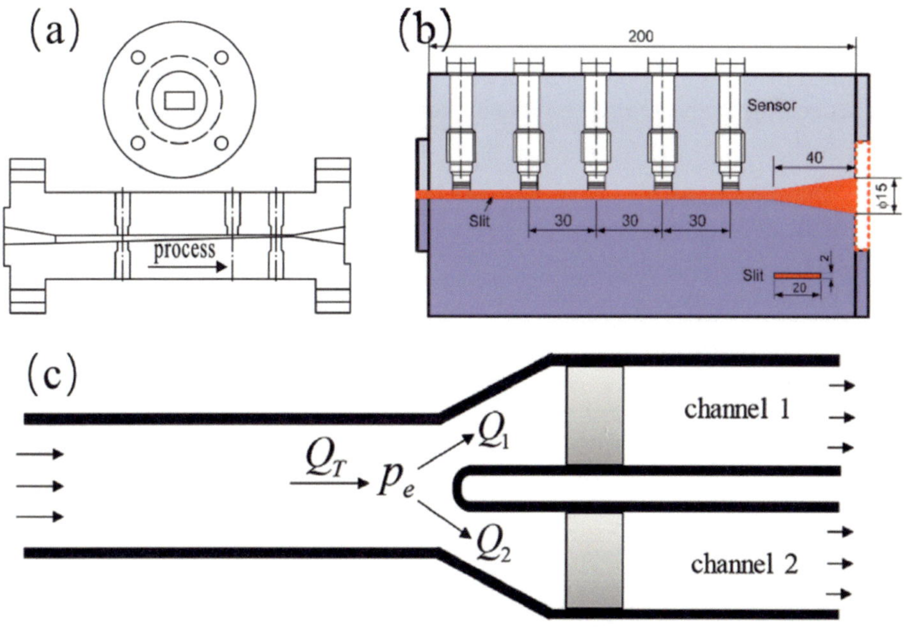

Figure 7.
Different slit rheometers: (a) online slit rheometer installed in extruder [35]; (b) online slit rheometer installed in injection molding machine [36]; and (c) inline slit rheometer with two channels [37].

is changed to a certain extent due to its processing thermal history being different from the actual mainstream channel, which is the common disadvantage of the online slit rheometer and the commercial offline rheometer. While the online rheometer has undergone significant enhancements compared to its offline counterpart, the outcomes still deviate from those observed in the mainstream channel. (2) The installation of a slit rheometer at the exit position poses an impediment to the regular processing of the material. Consequently, it is predominantly employed in research and development settings, utilizing a bypass for testing. (3) As viscosity requires data at multiple shear rates, different processing conditions are necessary. The thermal stability, homogeneity, and morphology of the material are affected.

Real-time capillary rheometers can be divided into inline and online types. Early developed capillary rheometers are similar to the online slit rheometers, and their installation position is also divided into the mold head position and the main flow channel bypass position [38], as shown in **Figure 8(a)**. In addition to bypass, real-time capillary rheometers installed at the injection molding machine and extruder die position also develop gradually. Limper [39] developed an online capillary rheometer to solve the problem that commercial offline rheometers cannot measure the rheological properties of PVC materials, which is used for the online determination of the flow characteristics of PVC materials. The drawback is that the power of the rheometer needs to be bigger, and the range of the achievable shear rate is only between 10/s and 200/s, which limits the wide application of flowmeters. Similarly, in the field of injection molding, Lohr [29] devised an online capillary rheometer integrated into an injection mold, enabling the concurrent assessment of flow characteristics for both the materials within the mold and the injection-molded products.

Compared with the online capillary rheometer, there are very few reports on the inline type. In 2022, Wappler [32] developed an inline capillary rheometer that can be applied to injection molding machines. As shown in **Figure 8(b)**, the measurement area is located between the plasticizing area of the barrel and the injection nozzle. The diameter and length of the capillary test section in the main flow path are initially determined through computer simulations, leading to variations in capillary dimensions depending on the specific melt being tested. Wappler et al. designed a capillary

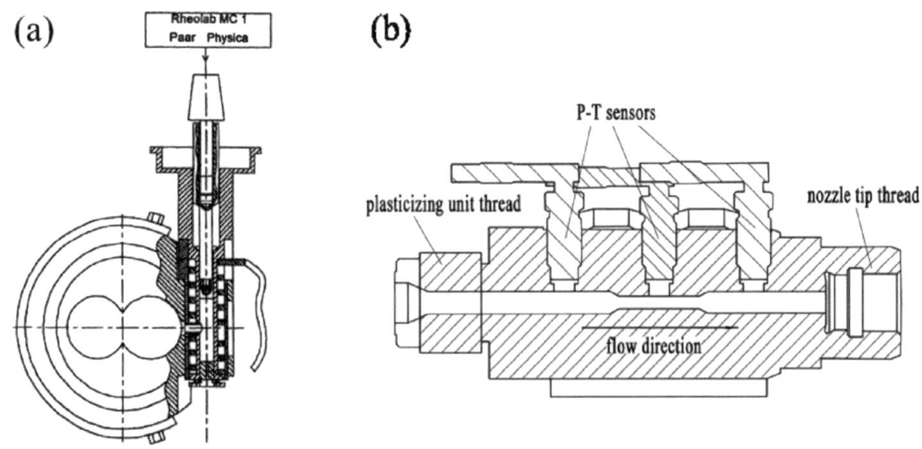

Figure 8.
Different kinds of real-time capillary rheometer: (a) online cap capillary rheometer [38]; (b) inline cap capillary rheometer [32].

smaller than the main channel but significantly more extensive than a conventional capillary rheometer's test area. This modification reduces pressure loss due to sudden deformation at the capillary inlet, a common issue in conventional capillary rheometers that necessitates Bagley correction. Additionally, because the measurement area is integrated into the main flow path, this design enables precise measurement of material flow characteristics. However, this design has a limitation: adapting the system to different units is difficult, as the capillary's aspect ratio is fixed. Researchers must calculate the required aspect ratio based on the specific melt material and experimental conditions before setting up the instrument. The need for different capillaries with varying aspect ratios for different polymer materials represents a significant challenge, potentially limiting the widespread adoption of this type of online rheometer.

4.2 Real-time rotation rheometer

Real-time rotational rheometers are generally flat plate rheometers and are only available online. The device is mounted on the bypass of the main flow path. Utilizing upper and lower plates, low-speed shear is applied to the material between the plates, and the torque transmitted by the two plates provides feedback on the material's rheological properties.

Mold et al. [4] developed an online rotational rheometer for twin-screw extruders. This device switches a valve on and off to take melt from the main flow path for rotational measurements. It has the same functionality as commercial rheometers but lacks thermal history from the valve to the rheometer. **Figure 4** describes the detailed process of the online rotational rheometer test.

4.3 Real-time oscillatory rheometer

Real-time oscillatory rheometers are used for dynamic viscoelastic testing of materials and can be worked online. Alternating stresses are applied to the material utilizing upper and lower plates, and the data collected by the sensors are used to calculate the energy storage modulus, loss modulus, and loss angle of the material. Konigsberg [40] incorporated an online oscillatory rheometer into the pipeline production process to test the rheology of the non-Newtonian fluid inside the pipeline, as shown in **Figure 9**. Although inline dynamic viscoelastic testing is possible, it is limited and only suitable for large pipeline production. It cannot be applied to processes such as extrusion and injection molding.

4.4 Real-time ultrasonic rheometer

In addition to the above rea-time rheometers based on traditional principles, several testing techniques based on new principles have emerged. Ultrasonic rheometers are ideally suited to measure the rheological properties of materials due to their non-destructive properties. Ultrasonic rheometers work on receiving the backscattered sound emitted by each pulse modulation, performing quadrature demodulation and ranging, obtaining the power spectrum, establishing the relationship between power and flow rate, and thus obtaining the material's flow characteristics. Pfund's research [41] team has measured the flow characteristics of non-Newtonian fluids using the patented ultrasonic Doppler velocimeter (UDV)-based inline rheometer, which obtains rheograms of shear-thinning gel. The study employed pulsed Doppler ultrasonic technology to measure the fluid's velocity profile within a pipe, enabling

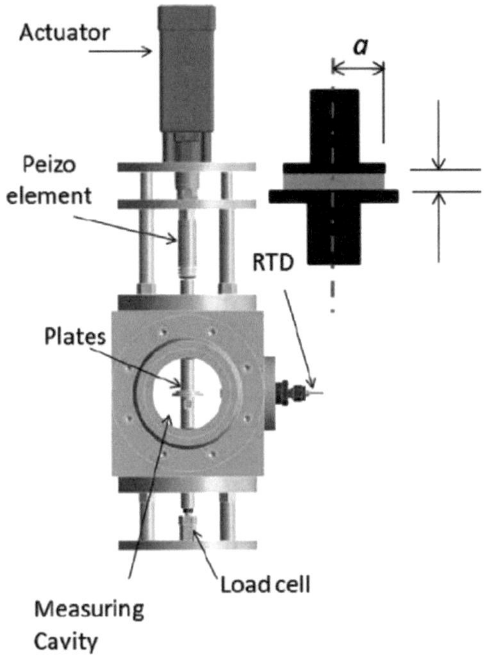

Figure 9.
Online oscillatory rheometers [40].

the estimation of rheological parameters, such as viscosity, from the acquired velocity data. This technique is based on the interaction of ultrasonic signals with scatterers present in the fluid, including particles, bubbles, and other inhomogeneities.

5. Real-time rheological testing applications

Polymers are essential to non-Newton materials, so real-time rheological testing technology has been widely applied in polymer processing and quality control. As previously mentioned, the primary application scenarios for the discussed concept involve extrusion and injection molding processes.

Pabedinskas [35] fabricated an online slit rheometer used in polymer extrusion area. An increasing number of scholars are engaged in advancing inline rheometers for applications in the extrusion process and injection molding. Kopplmayr [42] developed an online rheometer for extrusion and measured tensile and shear rheology simultaneously using hyperbolic slits. Sousa [43] applied the inline rheometer on a micro-extruder to produce 3D printing filaments, expanding the use of inline rheometers to a certain extent. An online rheometer with a hyperbolic slit contraction [8], as described by Unger et al. [44], was employed to measure the shear and extensional viscosity of the prepared PP/UHMWPE mixture. This approach was utilized to enhance the machining properties of UHMWPE and investigate the mixture's rheological behavior and mechanical property variations.

Volpe [45] employed an inline slit rheometer installed at the nozzle of an injection molding machine to achieve high shear rate rheological characterization and widened the range of inline rheometer shear rate testing for injection molding machines to 10^5 S^{-1}. Gou [46] proposed a novel real-time rheological measurement device integrated into the head of an injection molding machine, featuring two capillary tubes with different aspect ratios for Bagley correction. The experimental results indicate that real-time rheological testing more accurately reflects the actual machining environment, providing more precise rheological data, which is beneficial for enhancing the accuracy of numerical simulations and the optimization of machining parameters. Fernandez et al. [47] investigated the rheological behavior of polypropylene containing 10% mineral filler during injection molding using an inline rheological testing method. Moreover, the modified method can be applied to CAE simulations of non-traditional injection molding processes, enhancing mold design accuracy and optimization efficiency. Zhang et al. [48] developed an online rheological measurement system based on a dumbbell-shaped slit die to study the microinjection molding process. The results demonstrate that slit thickness, wall slip, and non-isothermal conditions significantly impact viscosity. Coogan [49] designed a novel 3D printer nozzle with a pressure sensor and thermocouple. It can monitor volumetric flow and shear rate by tracking stepper motor pulses, as well as custom wire encoder signals to account for wire slip and motor stepping errors. The inline rheometers deliver accurate viscosity measurements with appropriate corrections, facilitating real-time monitoring and process control.

The examples above suggest that using real-time rheometers to guide and control polymer processing is the way forward. In the non-plastic processing fields, such as the food and beverage industry, real-time rheometers also have a wide range of applications. The materials used in the food and beverage field mainly belong to non-Newton fluid materials; the rheological properties of such fluids often affect the efficiency of the processing operations and the quality of the products [40]. Unlike the high-temperature conditions of polymer processing, the rheology properties of these materials were measured under temperature around room temperature. Another important distinction is the higher requirements for the processing environment; these differences bring about different design requirements and structures. For example, for the monitoring and control of opaque non-Newtonian fluids, a combination of ultrasonic Doppler velocity profile (UVP) and differential pressure (PD) techniques (PUV + PD) is employed to observe the evolution of velocity profiles, the formation of mixing zones, and real-time variations in fluid rheology [50].

6. Challenges and future perspectives

Although real-time rheological measurement technology has been utilized in the field of plastics, yielding significant benefits, several challenges persist despite ongoing advancements in research. Firstly, commercially available and custom-made real-time rheological test instruments share a common measurement principle. However, their internal structures exhibit variations, lacking a standardized configuration in the polymer rheological testing process. Furthermore, certain online testing processes necessitate manual intervention, such as the post-testing cleanup of waste generated in some online capillary rheometers. This manual aspect hinders the attainment of a high degree of automation, posing challenges to industrial automation production. In the trajectory toward industrialization, enhancing the uniformity of equipment involved in real-time measurement technology is imperative. Simultaneously, efforts should be directed at augmenting the automation level of the testing process to minimize the impact of real-time rheological testing on polymer processing. Secondly, the rheological parameters of a given material are susceptible to influences from shear rate, shear stress, and temperature. The installation of instruments introduces variables such as slit size changes and differing distances between the main flow channel and bypass in various instruments. These structural factors contribute to disparities in measurement outcomes among real-time rheological testing instruments. Therefore, the structural design of the instrument should prioritize easy installation and testing, minimizing the impact of sampling on mainstream production and ensuring consistency in material properties between the bypass and the mainstream. When selecting sensors, preference should be given to those with higher precision and superior anti-interference capabilities, and careful consideration should be given to the sensor's installation position. Finally, researchers aiming to ensure the reliability of test results should choose different types of rheological test instruments based on the specific rheological properties of the materials under examination.

With advancements in measurement and testing methods, the aforementioned challenges will be gradually addressed, leading to a more robust and comprehensive rheological measurement system. To facilitate the widespread adoption of real-time rheological measurement in the plastics industry, future development directions for these techniques are as follows: (1) higher level of automation, (2) excellent adaptability to different equipment, and (3) non-destructive real-time rheological measuring techniques. Intelligent real-time rheometers will be developed and then applied in the field of polymer processing and quality control, thus promoting the development of the polymer processing industry and, at the same time, improving the real-time rheological testing technology.

7. Summary

This review mainly described real-time rheological testing techniques, equipment, and application areas in the plastics industry as thoroughly as possible. Existing real-time rheological test technologies were classified into online and inline based on their installation form, and the basis for their classification was clarified. Through different testing principles, the real-time rheological testing technology was divided into real-time slit and capillary rheometer, real-time rotational rheometer, real-time oscillatory rheometer, and real-time ultrasonic rheometer. The working principle of each real-time rheological measurement technology was introduced. The development of various types of real-time rheometers was also elaborated, which introduced the researcher's improvement and ingenuity for each type of real-time rheometers in terms of structure, principle, and other aspects. Subsequently, the application areas of real-time rheometers were introduced. Some implementation cases of real-time rheological measurement techniques applied in polymer processing and quality control, the food and beverage industry, the pharmaceutical and cosmetic industry, and the biomedical field were illustrated. Finally, the challenges and future perspectives of real-time rheometers were summarized.

Acknowledgements

This research was funded by the Fundamental Research Funds for the Central Universities of China (grant number buctrc202005).

Conflict of interest

The authors declare no conflict of interest.

Author details

Yucan Liang[1,2], Yue Cheng[1,2] and Jian Wang[1,2*]

1 State Key Laboratory of Organic-Inorganic Composites, Beijing University of Chemical Technology, Beijing, China

2 College of Mechanical and Electrical Engineering, Beijing University of Chemical Technology, Beijing, China

*Address all correspondence to: wjj_0107@163.com

References

[1] Aho J, Boetker JP, Baldursdottir S, Rantanen J. Rheology as a tool for evaluation of melt processability of innovative dosage forms. International Journal of Pharmaceutics. 2015;**494**(2): 623-642

[2] Baum M, Anders D, Reinicke T. Approaches for numerical modeling and simulation of the filling phase in injection molding: A review. Polymers. 2023;**15**(21):4220

[3] Krishnan JM, Deshpande AP, Kumar PS, editors. Rheology of Complex Fluids. New York: Springer; 2010. pp. 3-34

[4] Mould S, Barbas J, Machado AV, Nóbrega JM, Covas JA. Measuring the rheological properties of polymer melts with on-line rotational rheometry. Polymer Testing. 2011;**30**(6):602-610

[5] Bean RH, Rau DA, Williams CB, Long TE. Rheology guiding the design and printability of aqueous colloidal composites for additive manufacturing. Vinyl Additive Technology. Jul 2023;**29**(4):607-616

[6] Das A, Gilmer EL, Biria S, Bortner MJ. Importance of polymer rheology on material extrusion additive manufacturing: Correlating process physics to print properties. ACS Applied Polymer Materials. 2021;**3**(3):1218-1249

[7] Chen JY, Yang KJ, Huang MS. Online quality monitoring of molten resin in injection molding. International Journal of Heat and Mass Transfer. 2018;**122**:681-693

[8] Luger HJ, Miethlinger J. Development of an online rheometer for simultaneous measurement of shear and extensional viscosity during the polymer extrusion process. Polymer Testing. 2019;**77**:105914

[9] Klozinski A, Barczewski M. Comparison of off-line, on-line and in-line measuring techniques used for determining the rheological characteristics of polyethylene composites with calcium carbonate. Polimery. 2019;**64**(2):83-92

[10] Hencky H. Welche Umstände bedingen die Verfestigung bei der bildsamen Verformung von festen isotropen Körpern? Zeitschrift für Physik. 1929;**55**(3-4):145-155

[11] Ostwald W. Zur theorie der Liesegang'schen Ringe. Kolloid-Zeitschrift. 1925;**36**:380-390

[12] Ostwald W. Colloids and their viscosity. A general discussion. Transactions of the Faraday Society. 1913;**9**:34-46

[13] Sangroniz L, Fernández M, Santamaria A. Polymers and rheology: A tale of give and take. Polymer. Apr 2023;**271**:125811

[14] Onogi S, Masuda T, Kitagawa K. Rheological properties of anionic polystyrenes. I. Dynamic viscoelasticity of narrow-distribution polystyrenes. Macromolecules. 1970;**3**(2):109-116

[15] Abdel-Khalik SI, Hassager O, Bird RB. Prediction of melt elasticity from viscosity data. Polymer Engineering & Science. 1974;**14**(12):859-867

[16] Leaderman H. Creep and creep recovery in plasticized polyvinyl chloride. Industrial & Engineering Chemistry. 1943;**35**(3):374-378

[17] Williams ML, Landel RF, Ferry JD. The temperature dependence of relaxation mechanisms in amorphous polymers and other glass-forming liquids. Journal of the American Chemical Society. 1955;**77**(14):3701-3707

[18] Soares K, da Cunha Santos AM, Canevarolo SV. In-line rheo-polarimetry: A method to measure in real time the flow birefringence during polymer extrusion. Polymer Testing. 2011;**30**(8):848-855

[19] Teixeira PF, Hilliou L, Covas JA, Maia JM. Assessing the practical utility of the hole-pressure method for the in-line rheological characterization of polymer melts. Rheologica Acta. 2013;**52**:661-672

[20] Silva J, Santos AC, Canevarolo SV. In-line monitoring flow in an extruder die by rheo-optics. Polymer Testing. 2015;**41**:63-72

[21] Covas JA, Carneiro OS, Costa P, Machado AV, Maia JM. Online monitoring techniques for studying evolution of physical, rheological and chemical effects along the extruder. Plastics, Rubber and Composites. 2004;**33**(1):55-61

[22] Schaible T, Bonten C. In-line measurement and modeling of temperature, pressure, and blowing agent dependent viscosity of polymer melts. Applied Rheology. 2022;**32**(1):69-82

[23] Covas JA, Maia JM, Machado AV, Costa P. On-line rotational rheometry for extrusion and compounding operations. Journal of Non-Newtonian Fluid Mechanics. 2008;**148**(1-3):88-96

[24] Schweizer T, Schmidheiny W. A cone-partitioned plate rheometer cell with three partitions (CPP3) to determine shear stress and both normal stress differences for small quantities of polymeric fluids. Journal of Rheology. 2013;**57**(3):841-856

[25] Wasserman SH, Graessley WW. Effects of polydispersity on linear viscoelasticity in entangled polymer melts. Journal of Rheology. 1992;**36**(4):543-572

[26] Hilliou L, Covas JA. In-process rheological monitoring of extrusion-based polymer processes. Polymer International. 2021;**70**(1):24-33

[27] Mould ST, Barbas JM, Nóbrega JM, Machado AV, Covas JA. Recent developments on on-line Rheometry to monitor the extrusion process. In: Zlin (Czech Republic), AIP Conference Proceedings. Vol. 1152. No. 1. American Institute of Physics; 2009. pp. 175-182

[28] Mazzanti V, Mollica F. In-line rheometry of polypropylene based wood polymer composites. Polymer Testing. 2015;**47**:30-35

[29] Lohr C, Dieterle S, Menrath A, Weidenmann KA, Elsner P. Rheological studies on gas-laden and long glass fiber reinforced polypropylene through an inline high pressure capillary rheometer in the injection molding process. Polymer Testing. 2018;**71**:27-31

[30] Qwist PK, Sander C, Okkels F, Jessen V, Baldursdottir S, Rantanen J. On-line rheological characterization of semi-solid formulations. European Journal of Pharmaceutical Sciences. 2019;**128**:36-42

[31] Bonino CA, Samorezov JE, Jeon O, Alsberg E, Khan SA. Real-time in situ rheology of alginate hydrogel photocrosslinking. Soft Matter. 2011;7(24):11510-11517

[32] Wappler P, Horter T, Kulkarni R, Guenther T, Fritz KP, Zimmermann A.

Development of a nozzle capillary viscometer for inline viscosity measurement of thermoplastics. The International Journal of Advanced Manufacturing Technology. 2022;**122**(1):105-116

[33] Kotzé R, Wiklund J, Haldenwang R. Application of ultrasound Doppler technique for in-line rheological characterization and flow visualization of concentrated suspensions. The Canadian Journal of Chemical Engineering. 2016;**94**(6):1066-1075

[34] Kloziński A, Jakubowska P. The application of an extrusion slit die in the rheological measurements of polyethylene composites with calcium carbonate using an in-line rheometer. Polymer Engineering & Science. 2019;**59**(s2):E16-E24

[35] Pabedinskas A, Cluett WR, Balke ST. Development of an in-line rheometer suitable for reactive extrusion processes. Polymer Engineering & Science. 1991;**31**(5):365-375

[36] Moon JS, Lee JM. Shear viscosity measurement of highly filled polycarbonate melts using a slit-die rheometer. Korea-Australia Rheology Journal. 2013;**25**(3):129-135

[37] Teixeira PF, Ferrás LL, Hilliou L, Covas JA. A new double-slit rheometrical die for in-process characterization and extrusion of thermo-mechanically sensitive polymer systems. Polymer Testing. 2018;**66**:137-145

[38] Covas JA, Nóbrega JM, Maia JM. Rheological measurements along an extruder with an on-line capillary rheometer. Polymer Testing. 2000;**19**(2):165-176

[39] Limper A, Fattmann G, Seibel S. Development of a portable online rheometer for the characterization of PVC melts. Macromolecular Materials and Engineering. 2002;**287**(11):729-733

[40] Konigsberg D, Nicholson TM, Halley PJ, Kealy TJ, Bhattacharjee PK. Online process rheometry using oscillatory squeeze flow. Applied Rheology. 2013;**23**(3):35688

[41] Pfund DM, Greenwood MS, Bamberger JA, Pappas RA. Inline ultrasonic rheometry by pulsed Doppler. Ultrasonics. 2006;**44**:477-482

[42] Köpplmayr T, Luger HJ, Burzic I, Battisti MG, Perko L, Friesenbichler W, et al. A novel online rheometer for elongational viscosity measurement of polymer melts. Polymer Testing. 2016;**50**:208-215

[43] Sousa J, Teixeira PF, Hilliou L, Covas JA. Experimental validation of a micro-extrusion set-up with in-line rheometry for the production and monitoring of filaments for 3D-printing. Micromachines. 2023;**14**(8):1496

[44] Unger T, Klocke L, Herrington K, Miethlinger J. Investigation of the rheological and mechanical behavior of polypropylene/ultra-high molecular weight polyethylene compounds related to new online process control. Polymer Testing. 2020;**86**:106442

[45] Volpe V, Pantani R. Determination of the effect of pressure on viscosity at high shear rates by using an injection molding machine. Journal of Applied Polymer Science. 2018;**135**(24):45277

[46] Gou G, Xie P, Yang W, Ding Y. Online measurement of rheological properties of polypropylene based on an injection molding machine to simulate the injection-molding process. Polymer Testing. 2011;**30**:826-832

[47] Fernandez A, Muniesa M, Javierre C. In-line rheological testing of thermoplastics and a monitored device for an injection moulding machine: Application to raw and recycled polypropylene. Polymer Testing. 2014;**33**:107-115

[48] Zhang N, Gilchrist MD. Characterization of thermo-rheological behavior of polymer melts during the micro injection moulding process. Polymer Testing. 2012;**31**(6):748-758

[49] Coogan TJ, Kazmer DO. In-line rheological monitoring of fused deposition modeling. Journal of Rheology. 2024;**63**:141-155

[50] Wiklund J, Stading M, Trägårdh C. Monitoring liquid displacement of model and industrial fluids in pipes by in-line ultrasonic rheometry. Journal of Food Engineering. 2010;**99**(3):330-337

Section 2

Rheological Measurement in Viscoelastic Materials

Chapter 2

Exploring the Rheological Properties of Viscoelastic Materials

Rahul Sharma and Noor Jahan

Abstract

Between two of the extremes i.e., ideally viscous and ideally solid behaviors of the materials, there is a window depicting dynamic interplay between the viscous and elastic properties of the materials classified as viscoelastic behavior. Consideration of this blended behavior is crucial in many scientific, commercial, and biological applications. In this chapter, we will try to unveil the complexities of these materials by first understanding the basics of the viscoelasticity, discussing the relevance of various parameters such as Deborah number, Storage modulus, loss modulus etc., and various equations developed to model the viscoelastic response of such materials. A brief overview of the measurement procedures, various techniques employed to understand the realm of viscoelastic materials will also be under great focus. At last, the practical applications bridging the theoretical perceptions with the real world will also be elaborated in this chapter.

Keywords: viscoelasticity, fading memory, deborah number, burger model, magnetorheological fluids

1. Introduction

Materials, with their diverse properties, plays a crucial role in a wide range of applications that shape our world. Exploring the response of materials to stress and strain, from the fluidity of liquids to the rigidity of solids, is a journey to understand core of materials. This chapter embarks on a captivating exploration into the world of rheology, with a special focus on the dynamic characteristics of viscoelastic materials. From the everyday to the extraordinary, understanding how materials flow and deform is a key to unlocking innovation across diverse industries. Rheology, a special branch to study the changes in the material's outer dimensions and inner properties under the application of certain stress and strain, provides us the lens to observe these changes and hence differentiate materials for various applications [1]. The subject of viscoelastic rheological characteristics has seen several significant advancements that have greatly benefited humanity. The use of smart materials has improved their rheological properties, enabling better control of the devices in technologies such as suspension systems, haptics, prostheses, teleoperations, etc. [2] These materials include magnetorheological fluids and magnetorheological elastomers stabilized by surface functionalization. Sanjay et al. investigated the impact of replacing carbonyl iron particles with iron oxide nanoparticles that were deposited on graphene oxide layers. This led to increased sedimentation stability and improved viscoelastic rheological characteristics [3]. Avina et al. recently examined the impact of adding biopolymers—almonds, guar gum, etc.—to cement-based composites and discovered improvements in the materials' rheological characteristics [4]. By blending nanoparticles in the network of viscoelastic surfactants, Olga et al. have discovered an improvement in the rheological properties proving highly desirable features for real-world applications, especially for fracturing fluids in oil recovery [5]. Owing to such characteristics and applications, it is worth giving a thorough consideration to the rheological properties of viscoelastic materials.

In our everyday world, we encounter matter in two forms: solids and fluids (liquids and gases). Think of it like-viscous materials, like honey resisting flow when we try to stir them and elastic materials on the other side like waistbands, hand bands etc. which tends to regain their initial shape after the removal of the stress under which they were kept. Interesting fact is that there are some peculiar kind of materials which encompasses the properties of both of these extremes. These amazing materials are renowned as viscoelastic materials which serves as a bridge between purely elastic materials with strong memory and purely viscous materials being memoryless. These viscoelastic materials have a bit of both, a kind of fading memory [6]. These viscoelastic materials have the ability to change their behavior depending on the time with the applied shear stress. Toothpaste, Paints are the most common examples of these materials. Thus, viscoelastic materials have a time dependent behavior [7]. Now, to understand this intriguing behavior of such interesting materials we need some special techniques which can minutely and accurately tell us about the properties of these materials. Rheology, rheological techniques and analysis of these experiments serves this purpose. Understanding viscoelasticity is not just an academic pursuit—it holds the key to solving real-world problems in diverse fields. So, in this chapter, we embark on a journey to unravel the complexities of fluids that exhibit both viscous and elastic characteristics. This chapter will consist of various sections discussing the fundamentals of fluidic behavior, rheology of fluids, mathematical models such as Maxwell and Kelvin-Voigt, Oldroyd-B etc. providing the theoretical framework, toolkit for analysis of materials and their viscoelastic properties. With unfolding of each section, our understanding about the materials will get strengthened. Furthermore, we will discuss various parameters, experimental techniques used to characterize these materials and will delve deep into the process of unraveling the time-dependent nature of these viscoelastic materials. After that, various kind of fluids such as suspensions, colloids, polymeric fluids, biological systems will be under great focus. These viscoelastic materials have potential application in diverse fields including industry [8], healthcare [9], pharmaceutical [10] etc. each of them will be discussed in details. In the section that follows we will discuss the cases where the use of these viscoelastic materials have benefitted and in the end we will conclude the chapter with the key takeaways obtained from this exploration.

2. Fundamental of fluids and rheology

An intriguing question naturally arises: why delve into the fundamentals of fluids and rheology? The answer lies in establishing a foundational understanding that empowers readers of all expertise levels to navigate the mechanics and science governing these fluids. In the world of rheology, shear stress (σ), shear rate ($\dot{\gamma}$), and viscosity (η) are the dynamic trio that unveils the mesmerizing dance of fluids. Our exploration begins by probing the fundamental concepts such as shear stress, shear rate, and viscosity. This knowledge not only enhances the comprehension of our journey for readers in related fields but also serves as a gateway for a broader audience. Let us start by defining the stress acting on a material which can be basically of two types: Normal stress (σ) and shear stress (τ). Normal stress is the amount of stress that results from a force acting perpendicularly on a specific area in comparison to the shear stress which is the ratio of shearing force applied per unit area on a body with one end or base fixed. Shear stress tress is basically a quantity which means how much deforming force applies per unit area of an object. A type of stress that acts coplanar or parallel with cross section of material. It is a measure of how easily one plane within the material can be made to slide or deform along another in response to an applied force [11, 12]. Shear stress occurs when the material or fluid's adjacent layers move relative to one another due to an applied force. This results in a component of the force per unit area acting parallel to the surface. This force attempts to deform the material by causing one portion of it to slide past another. On the other side, shear rate or the rate of shear strain ($\dot{\gamma}$) is the rate at which shear deformation or strain (γ) is exerted on a particular material usually quantified as gradient of velocities. The most important parameter or property related to the fluid is the viscosity, the hidden dance partner of fluids which defines the resistance felt by fluids to flow. Just like the solids are identified by elasticity, fluids are identified with the help of viscosity. It's like the graceful choreography of a liquid, determining how smoothly or sluggishly it moves. It is the opposition that every single layer of the fluid offers to each other when they move relative to each other [12]. Thus, viscosity is the measure of the pressure required to cause a liquid to flow. Together, these three fundamentals define how different fluids behaves in term of their rheological properties, letting us distinguish between different fluids, unravel the mystery of their behavior. Let us now introduce several special characters that exist in the field of fluids: Newtonian fluids, non-Newtonian fluids such as Bingham plastic, pseudoplastic fluids, dilatant, and Herschel-Bulkley fluids. Newtonian fluids are the one following Sir Isaac Newton's law of viscosity strictly i.e., they have a constant viscosity regardless of shear rate or we can say that shear stress is directly proportional to shear rate [13].

$$\tau \propto \frac{d\gamma}{dt} \tag{1}$$

$$\tau = \eta \frac{d\gamma}{dt} \tag{2}$$

Where η represents the viscosity of the fluid. On the other hand, non-Newtonian fluids shows varying viscosity under different shear rates. They do not follow Newton's law of viscosity but tend to obey power law [14]. The general equation representing non-Newtonian fluids is,

$$\tau = \tau_y + m\left(\frac{d\gamma}{dt}\right)^n \tag{3}$$

Where yield stress(τ_y), is minimum initial stress required for the fluids to flow, m is consistency index telling us the thickness of the fluid and n is the flow index. This Eq. (3) is capable of explaining almost all type of fluids. For example, Dilatant fluids ($\tau_y = 0, n > 1$) thicken when you try to stir them quickly i.e., their viscosity increases

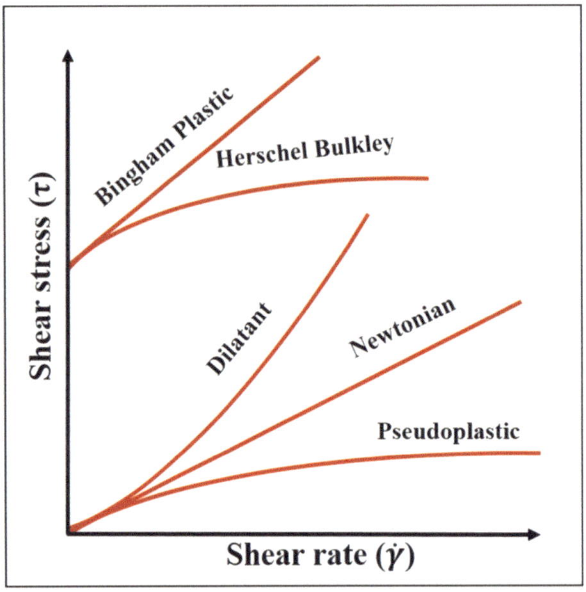

Figure 1.
Different types of fluids (redrawn from [21]).

non-linearly with the shear rate. Pseudoplastic fluids ($\tau_y = 0, n < 1$) on the other side, loses their viscosity upon increasing shear rate meaning they tend to become thin as the shear rate increases [15]. Yield stress τ_y comes into picture only in case of Bingham Plastics and Herschel-Bulkley fluids which starts to flow only after a certain value of stress applied to them called as "Yield stress" [16]. Bingham fluids follows the Newton's law of viscosity after this yield stress value is achieved [17, 18] whereas Herschel-Bulkley fluids shows Non-Newtonian behavior once they are exerted a stress value greater than the yield stress [19, 20]. Behavior of all these fluids is as shown in the **Figure 1**.

Armed with the essential information needed, we can immerse ourselves in the captivating realm of rheology and begin our exploration of viscoelasticity.

3. Basics of viscoelasticity

Viscoelastic materials possess a distinctive combination of elastic (solid-like) and viscous (liquid-like) properties, encompassing both the flowing properties of fluids and the resilient qualities of solids. However, what sets viscoelastic materials apart is their dual nature. They not only exhibit elastic behavior with the ability to recover from deformation but also display viscous behavior by flowing over time, resembling fluids. Examples include: polymers, some biological fluids, magnetic fluids etc. Understanding and modifying this coordination between the elastic and viscous aspects of these viscoelastic materials are crucial in diverse fields such as material science, engineering and biomechanics. Best example to explain the behavior of the viscoelastic materials are polymers. Toothpaste and paint etc. have polymers in them which cause them to show viscoelastic nature. Toothpaste consists polyethylene glycol (PEG), acrylic polymers are most commonly used in paints. In storage, paint should exhibit solid-like behavior to prevent sedimentation. However, during application, it should transform into a fluid-like state, allowing it to flow through the bristles of the brush for an even coat.

Origin of such behaviors having mixed characteristics can be understood from the entangled structure of the polymers that make up materials like paint and toothpaste. When polymers are highly entangled, they tend to show dominant elastic behavior with less fluidic characteristics, on the other flip, when disentanglement of polymers happens in these materials, they takes on a more viscous character behaving like fluids up to a large extent with minutely solid-like characteristics [22].

Viscoelastic materials exhibit time-dependent behavior due to their combination of viscous and elastic properties [23]. When stress is applied, these materials initially deform and then continue to respond over time. The extent and rate of deformation may vary, and factors such as temperature, frequency of applied stress or strain, and the specific viscoelastic properties of the material influence this time-dependent response which is extremely useful in tons of applications. When it comes to these viscoelastic materials, the stress at any given point in time depends not only on the current deformation gradient value but also on the entire temporal prehistory of the motion [24]. Viscoelastic substances exhibit diverse responses depending on the speed at which they undergo stretching. Note that the extent of displacement correlates with strain, making rate of strain the determinant of how rapidly the material experiences stretching. Therefore, viscoelastic materials are said to be strain rate-dependent.

Viscoelasticity can be linear or non-linear depending upon how the materials deform under the applied stress over a period of time. Viscoelasticity is categorized in three classes:

1. Linear viscoelasticity

2. Quasi- Linear viscoelasticity

3. Non-Linear viscoelasticity

Linear viscoelasticity deals with the behavior of viscoelastic materials having a linear relation between stress, strain at any instant of time. This means that if you double the stress or strain, the material's response will also double as seen in the **Figure 2**. It is based on Boltzmann's superposition principle which states that each loading step independently contributes to the final state. The well-established field of linear viscoelasticity applies to materials undergoing small deformations, such as the short-term deformation of polymer components. This concept of linear viscoelasticity came under focus Maxwell proposed the eq. [25]:

$$\frac{d\sigma}{dt} = E\frac{d\gamma}{dt} - \frac{\sigma}{\eta/_E} \tag{4}$$

Where, E is the Elastic modulus representing the stiffness of the material, η is the viscosity of the fluid filled in the dashpot. The ratio of η and E is called as the relaxation time usually denoted by λ. Over the time many scientists developed models to explain this particular behavior of materials but the initial thrust to this particular kind of viscoelastic behavior came from the Maxwell's equation. The following Boltzmann integral represents the linear viscoelasticity is as:

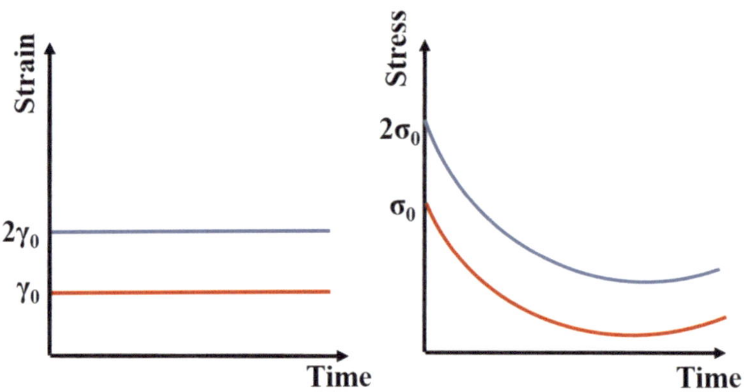

Figure 2.
Typical linear viscoelastic behavior of materials.

$$\sigma(t) = \int_0^t E(t-\tau)\frac{d\gamma}{d\tau}d\tau \tag{5}$$

Nonlinear viscoelasticity describes materials where the relationship between stress, strain, and time is not proportional (**Figure 3**). The material exhibits a nonlinear response, meaning that small changes in stress or strain can lead to disproportionately large changes in the material's behavior given by the integral:

$$\sigma(t) = \int_0^t E(t-\tau,\gamma\,(t))\frac{d\gamma}{d\tau}d\tau \tag{6}$$

This is common in materials subjected to high deformations or complex loading conditions. Nonlinear viscoelasticity is relevant in situations where materials experience large deformations, high stress levels, or complex loading conditions. Applications include the study of polymers, biological tissues, and materials subjected to intense mechanical forces.

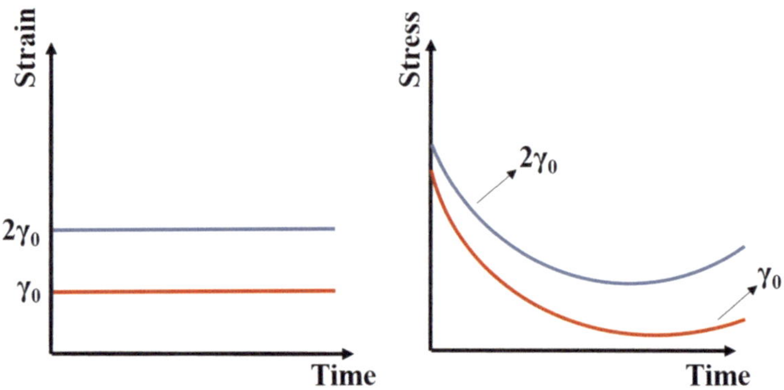

Figure 3.
Behavior of non-linear viscoelastic materials (red curve corresponds to the response of the material for γ_o strain whereas blue curve for strain $2\gamma_o$).

Quasilinear viscoelasticity bridges the gap between linear and nonlinear behavior. It allows for a certain degree of nonlinearity in the material response while retaining some linear characteristics. This model of quasilinear viscoelasticity was first presented by Y. C. Fung a long time ago as a straightforward method of combining nonlinearity and time dependence in a reduced integral model. This is particularly relevant when dealing with materials that may exhibit nonlinear behavior under specific conditions but still maintain some linearity in their overall response [26]. The equation which represents the Quasi-linear viscoelasticity is as:

$$\sigma(\gamma, t) = \int_0^t E_t(t-\tau)\frac{d\sigma}{d\gamma}\frac{d\gamma(t)}{d\tau}d\tau \tag{7}$$

In case of linear viscoelasticity, stress and strain exhibit a linear relationship, and notably, the relaxation modulus remains independent of strain. Despite the non-linear correlation between stress and strain, the relaxation modulus maintains its strain-invariant nature. In contrast, non-linear viscoelastic materials showcase a non-linear connection between stress and strain. For materials exhibiting linear or quasi-linear viscoelasticity, stress-versus-time graphs display a linear time dependence, and the shape of the graph remains consistent for different stress levels. However, in the case of non-linear viscoelastic materials, the time dependence in stress-versus-time graphs is non-linear, signifying a distinct behavior in response to applied stress [27].

4. Important parameters

Before dealing with these viscoelastic materials, we first must need to get familiar with certain important parameters like Deborah number, Weissenberg number, storage modulus, loss modulus etc. which serves as compasses guiding us through the intricate behaviors of the materials. To understand the mystery behind the behaviors of these materials that why they tend to behave in a certain fashion, when a particular material will behave in a similar fashion can be predicted using these parameters and that's why it becomes important to understand their significance.

4.1 Deborah number

Although relaxation time gives the information whether material will be solid or liquid by its value ranging from zero to infinity but when dealing with these materials, the time scale of the process needs to be considered. Deborah number, denoted as De, is a dimensionless parameter in the field of viscoelasticity that characterizes the ratio of the characteristic (relaxation) time of a materials response to an applied stress to the characteristic time scale of the deformation process [28]. It is represented as:

$$\mathrm{De} = \frac{\textit{Relaxation time}}{\textit{Observation time}} = \frac{\lambda}{t} \tag{8}$$

This unique parameter signifies the effect of time scale of the process on material's deformational behavior. "Fluid-like" behavior is to be predicted if the material is observed for an extended period of time or if it relaxes quickly (i.e. De $< <1$). On the other hand, if the material has a long relaxation time or a short observational period, the Deborah number is high (De > > 1) and the material behaves like a solid in all practical senses.

4.2 Weissenberg number

Weissenberg number is another parameter which is quite important while dealing with the viscoelastic materials. It is denoted by Wi and is equal to two times the product of shear rate $\dot{\gamma}$ and relaxation time (λ) [29].

$$Wi = \frac{Elastic\ force}{Viscous\ force} \tag{9}$$

$$Wi = \frac{2\lambda\eta\dot{\gamma}^2}{\eta\dot{\gamma}} = 2\dot{\gamma}\lambda \tag{10}$$

A low value of Wi elucidates the process to be linearly viscoelastic whereas high Wi number indicates non-linearly viscoelastic response of the material. It is an useful metric which quantifies the level of anisotropy or alignment caused by the displacement of the system from original position and is apt for describing flows with a consistent stretching background [30].

4.3 Storage modulus, loss modulus and loss tangent

To forecast the behavior, a viscoelastic material's modulus of elasticity needs to be able to account for both the material's elastic energy-storing and viscous energy-dissipating components. This is accomplished by calculating the complex modulus, represented by E^*, which is the total of the viscous component, represented by E'', and the elastic component, called storage modulus, denoted by E' [31].

$$E^* = E' + E'' \tag{11}$$

These crucial variables show up as the main guides for comprehending how the materials react when shear force is applied. The storage modulus provides insight into a material's resilience and ability to regain its original shape by indicating its ability to store energy elastically under deformation. It is the real part of the complex modulus (E^*). On the contrary, the imaginary component of the complex modulus, or loss modulus, indicates how much energy a material loses under stress. With little energy lost and nearly complete recovery, a material with a high shear storage modulus (E') and a low shear loss modulus (E'') will act more like a solid than a fluid. When E' is low and E'' is large, the material will mostly exhibit fluid behavior.

The loss tangent, also called the loss factor and represented by tanδ, is a more practical way to represent the viscoelastic material. It is the ratio of the loss modulus to the storage modulus [31].

$$tan\delta = \frac{E''}{E'} \tag{12}$$

When tanδ is higher than 1, the material mostly exhibits the characteristics of a viscous fluid. Whereas if it is less than one, the material will more solid like characteristics. Here, for most of the cases, we have considered the normal stress applied on materials therefore, we have considered modulus of elasticity (E), elastic storage modulus (E′) and loss modulus (E″) but for the case when a material is sheared with a coplanar force in that case, we take into account shear modulus (G), shear storage modulus (G′), shear loss modulus (G″). In that particular situation, we will have complex shear storage modulus (G*) as sum of G′ and G″ and loss tangent equal to the ratio of G″ and G′

5. Mathematical rheological models

Mathematical models serve as indispensable tools to bridge the gap between theoretical understanding and real-world applications. This section delves into the exploration of mathematical models employed to describe both linear and nonlinear viscoelasticity. Maxwell, Kelvin-Voigt, Burger model etc. are used to describe the linear viscoelastic behavior [32, 33] whereas non-linear intricacies inherent in materials when subjected to significant deformations or complex loading conditions are explored using the mathematical models such as Oldroyd-B and Wagner model etc.

5.1 Constitutive models for linear viscoelasticity

Behavior of these viscoelastic materials has been explained using the different arrangements of spring and dashpots. The various models which explain the linear behavior of viscoelasticity are

5.1.1 Maxwell model

Maxwell model of linear viscoelasticity consists of a spring and a dashpot attached in series arrangement [22] as shown in the **Figure 4**. The spring in the Maxwell model represents the elastic component of the material following the Hooke's law. In the context of the Maxwell model, this elastic element accounts for the instantaneous deformation of the material when stress is applied. The dashpot represents the viscous component of the material. It introduces a resistance to deformation proportional to the rate of change of deformation. In other words, the dashpot resists sudden changes in shape and dissipates energy over time. This viscous element captures the delayed, time-dependent response of the material. The primary equation representing the Maxwell model is

$$\frac{d\tau}{dt} = E\frac{d\gamma}{dt} - \frac{\tau}{\eta/E} \tag{13}$$

This equation can be solved for different η/E values i.e., relaxation time (λ) which indicates the rate of stress relaxation and in this way, we can see the behavior of the material with respect to time. When a constant stress is applied, the strain has two components i.e., elastic akin to the attached spring which relaxes immediately when the stress is released and viscous one corresponding to the dashpot that increases over time as long as the stress persists. The total deformation in the material is the sum of elastic and viscous contributions.

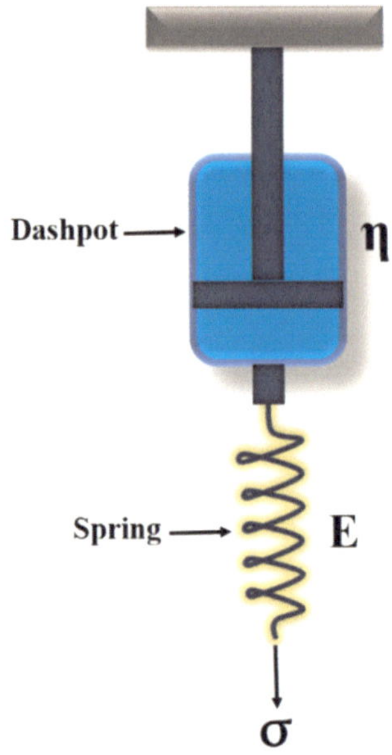

Figure 4.
Representation of a Maxwell model (this image is redrawn, drawing inspiration from [28]).

At constant strain (γ), strain rate is zero. So, from Eq. (13).

$$\frac{d\sigma}{dt} = -\frac{\sigma}{\lambda} \tag{14}$$

Now, by integrating this Eq. 14 and using the boundary condition ($\sigma = \sigma_m$), we have.

$$\sigma = \sigma_m \exp^{-t/\lambda} \tag{15}$$

Thus, subjecting the material to a constant strain result in the gradual relaxation of stress. At t = 0, $\sigma = \sigma_m$ which implies the response is of spring only i.e., elastic nature of the material whereas at t = λ, $\sigma = \sigma_m/e$ telling us that the stress dropped from its maximum value to σ_m/e. This model accurately predicts that the stress decay over time for many polymers but it falls short while explaining the creep behavior i.e., behavior of the material under constant stress. Under creep behavior, model predict linear increase in strain but in case of some polymers we have decreasing strain [34].

5.1.2 Kelvin-Voigt model

Another possible arrangement of elastic and viscous components i.e., a spring and dashpot or damper is their parallel combination (**Figure 5**) which describes the behavior of the materials under constant applied stress and provides the time dependent response of the materials [28]. In case of Maxwell model, the stress applied is same whereas the rate of strain is different. So, strain rates are added together to get total strain. In this model, since, there is parallel arrangement of spring and dashpot so the effect of force, strain rate will be same for both of the components but the stresses are different. So, the net stress will be the sum of stress

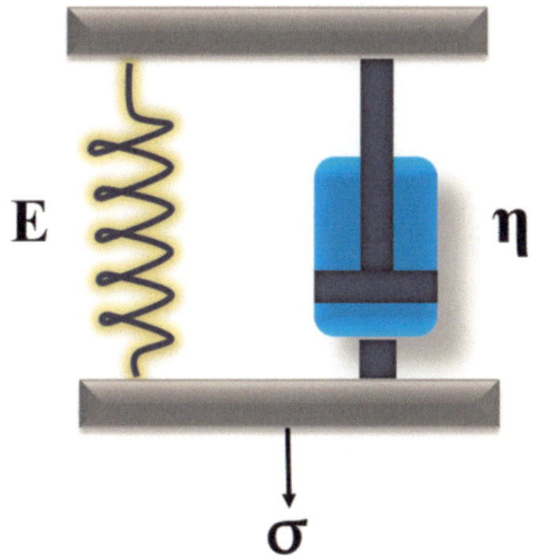

Figure 5.
Representation of a Kelvin-Voigt model (this image is redrawn, drawing inspiration from [28]).

induced in both of the components. The constitutive equation representing the behavior is as:

$$\sigma = E\gamma + \eta\dot{\gamma} \tag{16}$$

i.e.

$$\gamma = \frac{\sigma}{E} - \lambda\dot{\gamma} \tag{17}$$

At time t = 0, when a sudden force F is applied, we know dashpot cannot take instantaneous strain which means that there will be no strain on the spring part as they are in parallel. Initially dashpot carries the whole load. Now at time t, after a small period from t = 0, the dashpot will start to deform and therefore spring will also get deform. Hence, there will be equal deformation or strain in both the components. As the spring will stretch, there will be restoring force in the spring too. At some point of time say at t = ∞, the restoring force is equal to the applied force i.e., equilibrium position which also means that there will not be any force carried by the dashpot but only by the spring.

At t = ∞, elongation or deformation in spring will be,

$$\gamma_{\infty} = \frac{F}{k_s} \tag{18}$$

Now, the rate of deformation occurred in the dashpot will be,

$$\frac{d\gamma}{dt} = \frac{F_d}{k_v} = \frac{F - F_s}{k_v} \tag{19}$$

Now using Eq. (19), we can write,

$$\frac{d\gamma}{dt} = \frac{F - \gamma k_s}{k_v} = \left(\frac{k_s}{k_v}\right)\left(\frac{F}{K_s} - \gamma\right) \tag{20}$$

$$\frac{d\gamma}{dt} = \frac{F - \gamma k_s}{k_v} = \left(\frac{k_s}{k_v}\right)(\gamma_\infty - \gamma) \tag{21}$$

$$\frac{d\gamma}{dt} = \frac{(\gamma_\infty - \gamma)}{\lambda} \tag{22}$$

On using variable separable method for integrating above equation we get,

$$\gamma = \gamma_\infty\left(1 - e^{\frac{-t}{\lambda}}\right) \tag{23}$$

Because these elements are in parallel, they move together at a constant rate. Applying a sudden constant load to the Kelvin model initially yields delayed deformation, followed by a steady-state deformation. Upon load removal, the Kelvin model recovers completely but not instantaneously. The Kelvin-Voigt model can be used when a material exhibits both immediate responsiveness and persistent deformation.

Numerous other models have been created using different combinations of springs and dashpots, consisting of three and four elements. The Maxwell and Kelvin-Voigt models serve as the foundation for these three- and four-element models, which are constructed with various arrangements of dashpots and springs. One of the best of these is the burger model, which is covered in more detail below:

5.1.3 Burger model

One of the best 4-element models is Burger model. This is used to predict the creep behavior of number of materials. This model is obtained by placing the Maxwell and Kelvin-Voigt model in series with each other i.e., series combination of spring and dashpot is place in series with parallel combination of spring and dashpot [35, 36]. In such a combination (shown in **Figures 6** and 7), suspending a weight on the structure would cause the object weight or potential energy to distend the structure maximally and more slowly because of the damping effect of dashpot. Once the weight is removed, structure will slowly try to return to its original configuration due to the action of springs however due to the energy dissipated by the dashpots some of the strain is irrecoverable and the structure will not return to its original or baseline position. It is thus partially deformed permanently. There are two possible configurations representing this particular model, one given by Maxwell and other by Kelvin as

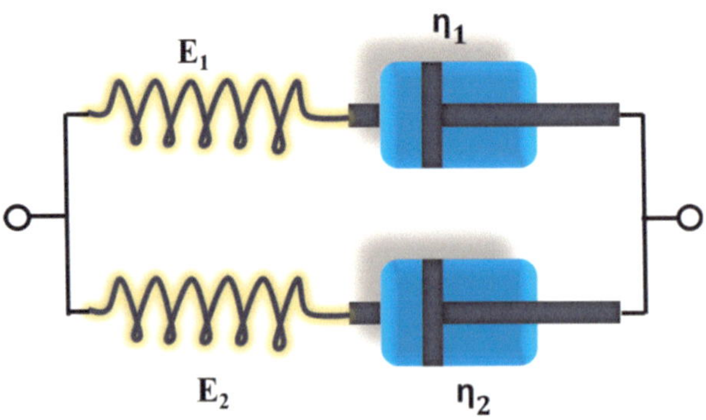

Figure 6.
Maxwell representation of Burger model (redrawn from the original source [37]).

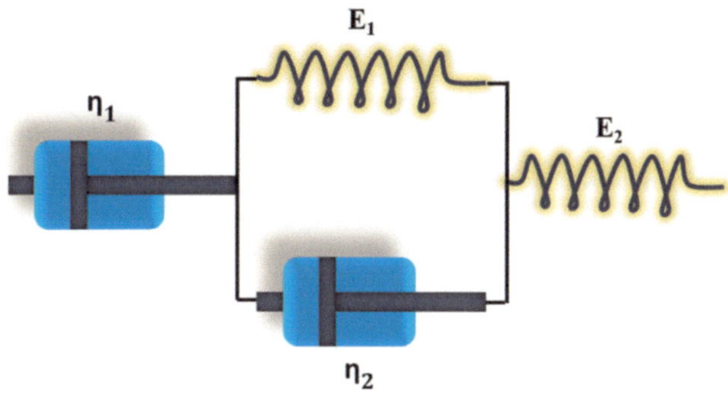

Figure 7.
Kelvin representation of Burger model (redrawn from the original source [37]).

shown in the Figures below. The Maxwell representation of constitutive equation for this particular model is given as:

$$\sigma + \left(\frac{\eta_1}{E_1} + \frac{\eta_2}{E_2}\right)\dot{\sigma} + \frac{\eta_1\eta_2}{E_1E_2}\ddot{\sigma} = (\eta_1 + \eta_2)\dot{\gamma} + \eta_1\eta_2\frac{(E_1 + E_2)}{E_1E_2}\ddot{\gamma} \tag{24}$$

Where, one Maxwell material has an elasticity E_1 and viscosity η_1, and the other Maxwell material has an elasticity E_2 and viscosity η_2.

The Kelvin representation of constitutive equation for this particular model is given as:

$$\sigma + \left(\frac{\eta_1}{E_1} + \frac{\eta_2}{E_2} + \frac{\eta_2}{E_1}\right)\dot{\sigma} + \frac{\eta_1\eta_2}{E_1E_2}\ddot{\sigma} = \eta_2\dot{\gamma} + \frac{\eta_1\eta_2}{E_1}\ddot{\gamma} \tag{25}$$

Where, one Kelvin material has an elasticity E_1 and viscosity η_1, and the other Kelvin material has an elasticity E_2 and viscosity η_2.

5.2 Constitutive models for non-linear viscoelasticity

To quantitatively address phenomena in fluids, such as stress variations, shear thinning and thickening, there is a requirement for non-linear viscoelastic constitutive equations. Various models or equations to account for this are discussed as below

5.2.1 Second order fluid model

The second-order fluid model is a constitutive model used to describe the rheological behavior of fluids, particularly those exhibiting non-Newtonian characteristics. Unlike Newtonian fluids, which have a constant viscosity, non-Newtonian fluids display a viscosity that varies with the applied shear rate or stress. The second-order fluid model is one approach to capturing the non-linear viscosity observed in certain complex fluids. The constitutive equation of this model is represented as:

$$T = -pI + 2\eta_0 D - \psi_{1,0}D^{\nabla} + 4\psi_{2,0}D.D \tag{26}$$

Where, T is the stress tensor, I is the identity tensor, D is the deformation tensor, $\eta_0, \psi_{1,0}$ *and* $\psi_{2,0}$ denote viscosity, and first and second normal stress coefficients, respectively and D^{∇} is the upper convected derivative of D [38]. The term proportional to D^{∇} in above equation incorporates a weak elastic "memory" into the constitutive equation. It can be shown under quite general conditions that a viscoelastic fluid will obey this above equation if the flow is sufficiently slow and slowly varying, to ensure that departures from Newtonian behavior are small. But this model is applicable to a limited number of non-Newtonian fluids only and does not cover a wide range of non-linear viscoelastic behaviors. This relation will hold only if the shear rate is low enough to prevent the viscosity and first and second normal stress coefficients from departing from their low shear rate values that is, if there is no shear thinning.

5.2.2 Upper-convected Maxwell model

The upper-convected Maxwell (UCM) model is a generalization of the Maxwell material for the case of large deformations using the upper-convected time derivative. The model was proposed by James G. Oldroyd. The concept is named after James Clerk Maxwell. Perhaps the simplest way to combine time-dependent phenomena and rheological nonlinearity is to incorporate nonlinearity into the simple Maxwell equation, This can be done by replacing the substantial time derivative in a tensor version of Maxwell equation with the upper-convected time derivative of τ [39]:

$$\tau + \lambda \tau^{\nabla} = 2\eta_0 D \tag{27}$$

This equation is the upper-convected Maxwell (UCM) equation, is nonlinear because τ^{∇} contains products of the velocity gradient ∇v and the stress tensor τ. For small strain amplitudes, the nonlinear terms disappear and the upper-convected time derivative reduces to the substantial time derivative UCM equation is then equivalent to the linear Maxwell model. On the other hand, if the flow is steady and the strain rate is small, τ is negligible and Newtonian behavior is recovered. Thus, to first order in the velocity gradient, we obtain.

$$\tau \approx 2\eta_0 D + \textit{second order terms} \tag{28}$$

Suppose that we now increase the strain rate until we start to see a weak departure from Newtonian behavior. We can calculate this departure by using the above equation to calculate upper convected stress tensor:

$$\tau^{\nabla} \approx 2\eta_0 D^{\nabla} + \textit{third order terms} \tag{29}$$

Then we get,

$$\tau \approx 2\eta_0 D - 2\eta_0 D^{\nabla} + \textit{second order terms} \tag{30}$$

Comparing this Eq. (30) with second-order fluid equation, we see that to second order in the velocity gradient the upper-convected Maxwell equation for small strain rates reduces to a special case of the equation of the second-order fluid with first order stress coefficient being $2\eta_0 D$ and second order stress coefficient being zero [40].

5.2.3 Oldroyd-B model

The Oldroyd-B model is an extension of the upper Convected Maxwell model [39] and is interpreted as a solvent filled with elastic bead and spring dumbbells. Although the UCM equation gives the polymer contribution to the stress in a dilute solution, the solvent contribution to the stress cannot be neglected, and so the total stress tensor in these solutions is the sum of the polymeric and solvent contributions

$$\tau = \tau^p + \tau^s \tag{31}$$

Where, τ^p is the polymeric contribution and is given by equation Upper Convected Maxwell model (eq. UCM model) and τ^s is the contribution due to solvent given by $2\eta_s D$ which ultimately gives,

$$\tau + \lambda_1 \tau^{\nabla} = 2\eta_0 \left(D + \lambda_2 D^{\nabla}\right) \tag{32}$$

Where, $\lambda_1$relaxation is time and λ_2 is retardation time $=\frac{\eta_s}{\eta_0}\lambda_1$. This model gives good approximations of viscoelastic fluids in shear flow, it has an unphysical singularity in extensional flow, where the dumbbells are infinitely stretched. This is, however, specific to idealized flow; in the case of a cross-slot geometry the extensional flow is not ideal, so the stress, although singular, remains integrable, although the stress is infinite in a correspondingly infinitely small region. If the solvent viscosity is zero, the Oldroyd-B becomes the Upper Convected Maxwell Model.

5.2.4 Wagner model

Wagner model is might be considered as a simplified practical form of the Bernstein–Kearsley–Zapas (BKZ) model [41, 42]. When the Cauchy strain terms are ignored in the original K-BKZ model we get Wagner model. This idea was first employed by Wagner [43, 44]. The Wagner model often involves a set of equations based on molecular reptation theory, describing the movement of polymer chains through entanglements. The specific equations can vary depending on the details of the model.

The constitutive equation for this model is given by,

$$\sigma = -p(\rho, T)I + \int_{-\infty}^{t} M(\xi(t) - \xi(t'))h(I_1, I_2)C_t^{-1}(t')dt' \tag{33}$$

Here, $M(\xi(t) - \xi(t'))$ the memory function carries the value,

$$M(\xi(t) - \xi(t')) = -\frac{dG(\xi(t) - \xi(t')}{dt} = \sum_{i=1}^{m} \frac{g_i}{a_T \theta_i} e^{-\frac{\xi(t)-\xi(t')}{\theta_i}} \tag{34}$$

And $G(\xi(t) - \xi(t') = \sum_{i=1}^{m} g_i e^{-\frac{\xi(t)-\xi(t')}{\theta_i}}$ is the shear relaxation modulus, θ_i is the relaxation time, $h(I_1, I_2)$ is the strain damping function.

6. Testing of viscoelastic materials

Several tests can be employed to investigate viscoelastic materials and establish the relation between stress, strain and time under specific deformation and loading conditions. The primary tests encompass stress relaxation, creep and dynamic tests. Rheometers which have been discussed in details in the initial chapters, are widely used by the rheologist to perform such tests on the materials under investigation to get the necessary information about their behaviors. These assessments are crucial for understanding the materials behavior and performance, providing valuable insights into its viscoelastic properties over time and under various loading patterns. Following section discusses these tests in detail.

6.1 Stress relaxation test

Stress relaxation test is a cornerstone in the study of viscoelastic materials offering a detailed exploration of their response to constant deformation or constant step strain over time. The primary aim of this test is to observe how a material gradually reduces stress under sustained strain, providing essential insights into its relaxation processes. Observing this relaxation of stress with time under the action of the constant deformation allows the observer to analyze the material's stress relieving capabilities and mechanism governing its viscoelastic properties. Maxwell model is commonly used to represent the stress relaxation behavior of viscoelastic materials. The experimental procedure involves subjecting a specimen to a constant deformation, with stress continuously monitored and recorded. The resulting stress decay curve, often exhibiting an exponential nature, serves as a key parameter for understanding the material's relaxation time and time dependent viscoelastic behavior.

A graph depicting stress versus time for a material stretched to a particular displacement and held there is shown in **Figure 8**. Vertically upward line at time t_0 is the force required to deform the material up to a certain stress say σ_0 on the application of

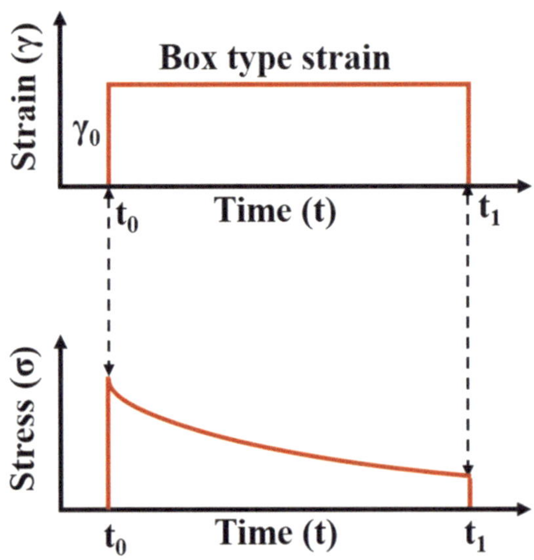

Figure 8.
Stress relaxation behavior of a viscoelastic material (this illustration has been redrawn using the concept from [22]).

a sudden strain or displacement, there is an increase in the stress induced in the material but when the strain is held constant further, there is a reduction in the induced stress with time. Approaching the end of the graph, the line flattens horizontally, indicating the attainment of a stable stress level and equilibrium within the material.

This stress relaxation after the application of a step strain (γ_0) is the fundamental way to define the relaxation modulus E (t), one of the most important viscoelastic parameters given as:

$$E(t) = \frac{\sigma(t)}{\gamma_0} \quad (35)$$

It is the time at which the stress in the body reduces to 1/e of its initial value. It signifies the rate at which the material releases the stress.

6.2 Creep and recovery test

Creep and recovery test provides crucial insight into how materials respond to a constant box type stress for some period of time as shown below. **Figure 9** shows the response of the viscoelastic material under such applied constant stress. Creep and recovery behavior of viscoelastic materials can be easily shown by the use of Burger model. This experiment involves application of a sudden load, holding it for some time and then removing the load.

It can be seen that at time t_0, there is an elastic strain produced in the material but in the interval from t_0 to t_1, strain is increasing always with the time such strain produced in the material under constant stress applied is called as creep strain. Now at time t_1, when the load is removed, the initial decrease in strain is equal to the amount of strain the material instantaneously experienced when it first had the force applied.

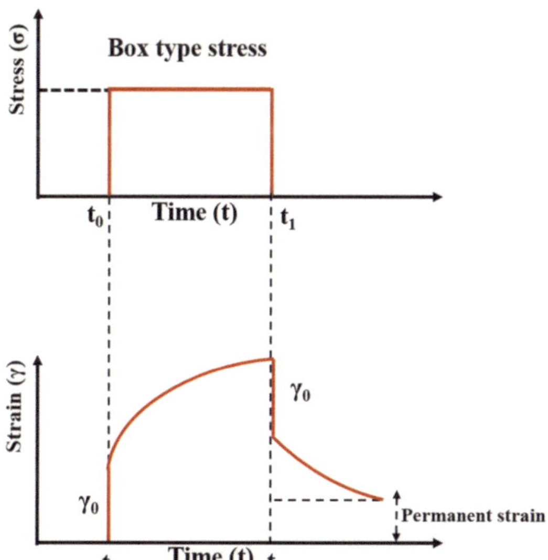

Figure 9.
Strain response of a viscoelastic material to the creep recovery test (this illustration has been redrawn using the concept from [22]).

Over time, the material almost returns to its original configuration and its strain becomes approximately zero with very small amount of inelastic recovery in case of metals but there is significant amount of unrecovered strain left in case of polymers [45].

6.3 Dynamic tests

Although stress relaxation and creep-recovery tests are quite simple techniques to characterize the viscoelastic materials but there are two disadvantages of these techniques, first is that we have to take measurements of a many decades of time cycle but during this time interval there can occur chemical and physical changes inside the material which can affect the behavior of the material and hence the final information that is extracted from the test. Second disadvantage is that is impossible to apply such instantaneous deformation at the beginning of the test. These issues are overcome with the help of dynamical or sinusoidal load when the material under investigation is loaded with a deformation varying sinusoidally with time. Common type of dynamic tests includes Dynamic Mechanical Analysis (DMA), vibration and oscillation testing etc. involving the exposure of specimen to varying loads or frequencies while measuring its dynamic response.

7. Practical applications

Due to their distinctive ability to deform and dissipate energy under stress, viscoelastic materials are extensively utilized across diverse industries. Their versatility renders them essential in numerous fields. This section digs into particular applications of viscoelastic materials, emphasizing their many applications and practical contributions. Various kind of viscoelastic applications are as discussed below (**Figure 10**).

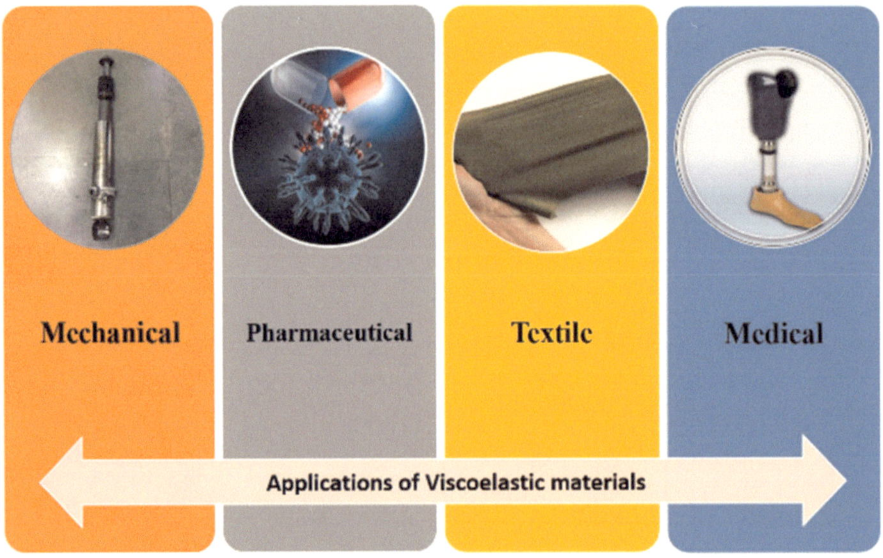

Figure 10.
Applications of viscoelasticity in various sectors (image source for pharmaceutical [46], textile [47] and medical applications [48]).

7.1 Mechanical applications

Viscoelastic materials finds their place in variety of applications ranging from enhancing structural resilience to improving the comfort and performance of every day products and machinery [49]. Blend of unique properties makes them invaluable in addressing various mechanical challenges across industries. These materials play a crucial role in the automotive sector, contributing to improved safety, comfort, and performance. They are extensively used in vehicle components such as shock absorbers, vibration dampers etc. These materials find substantial use in the manufacturing of sports equipment such as helmets, pads, and shoe soles. Their ability to absorb and dissipate impact energy helps prevent injuries by cushioning the effects of shocks and reducing the transmission of forces to the athlete. These viscoelastic materials and their coatings are very much useful in increasing the life span of the components of the machines by reducing the wear and tear issues. Viscoelastic dampers are used in the buildings and structures of the regions which are prone to earthquakes which reduces the effect of the vibrations in these structures by absorbing and then dissipating the effect thereby minimizing the potential damage. Viscoelastic materials are employed in the aerospace industry for interior components such as seating and panels. Their vibration damping capabilities contribute to passenger comfort by minimizing the effects of turbulence and mechanical vibrations during flight.

7.2 Pharmaceutical applications

Viscoelastic materials are now widely used in the pharmaceutical business [50], where they have a wide range of applications that help with drug discovery, production, dermatological uses, etc. [51]. These viscoelastic materials are frequently used in medicine formulation to improve the chemical and physical characteristics of the medication. Mucoadhesive and hydrogel polymers, for example, help to increase the stability, controlled release, and bioavailability of pharmaceuticals. Even viscoelastic encapsulation technologies are essential for shielding delicate medications from outside influences. Pharmaceutical companies use biodegradable viscoelastic polymers, particularly for designing implants or controlled drug release systems. Viscoelastic materials are used in dermatology to formulate lotions, ointments, and other products [52]. The therapeutic effects of these formulations are ensured by improving medication penetration through the skin.

7.3 In textile industries

Textile industry has witnessed a paradigm shift with the advent of the viscoelastic materials as they enable the development of the fabrics that meets the dynamic and diverse needs of modern consumers. When subjected to a load, textiles re recognized for exhibiting dynamic viscoelastic behavior [53]. These materials are known to produce hysteresis behavior under cyclic loading and unloading and they tend to release the stress when load is decreased with time [54]. These special qualities of the materials not only make the fabric flexible and comfortable, but they are also commonly utilized in sportswear to make it durable and able to withstand a lot of movements, which can greatly enhance a player's performance [55]. Impact collisions frequently result in numerous injuries to participants in physically hard-fought and competitive sports like football, rugby, boxing, etc. Use of the padding materials, safe guard mitigates the impact of such collisions by absorbing the forces up to 35 to 40%. D3O, Sorbothane (used in shoes) etc. are some of the materials used as an energy absorber in the sportswear [56].

7.4 Medical applications

As we are familiar with the fact that polymers are the representative for the viscoelastic materials. These materials holds a great position in the medical field [57]. Materials with viscoelastic properties can be employed as agents to deliver drugs into human bodies in order to treat specific illness [58, 59]. In addition to using them as drug delivery agents, we may use them to make medical equipment that can be used in the amputation of body parts. In the biomedical field, 3D printed and injection molded polymers are employed as prosthetics bearing products [60] emphasizing their applications in hip and knee joints [61]. Viscoelasticity is also observed in magnetic fluids especially magnetorheological fluids therefore they also hold a special place in the category of viscoelastic materials [62, 63]. Devices based on these magnetorheological fluids such as rehabilitation, haptic and prosthetic devices also plays an important usefulness in the field of medicine and surgery [64]. Prosthetic devices uses magnetorheological fluids based dampers for their controlled motion helping the amputees [65]. Thus, these magnetorheological viscoelastic materials are integral to the design of orthopedic implants and prosthetics. Viscoelastic materials are also used in ophthalmology as some viscoelastic materials are employed in the production of contact lenses to improve the comfort and to provide the stable fit on the eye. Moreover, due to their unique properties based on their chemical structure, these materials can also be used in the protection of corneal endothelium and epithelium from mechanical trauma and to maintain an intraocular space. Healon, Vitrax, Orcolon etc. are some of the viscoelastic materials that have been used earlier in ophthalmology [66]. In conclusion, the medical applications of viscoelastic materials encompass a broad spectrum of interventions and versatility of these materials contributes significantly to the advancement of medical treatments.

There are tons of other important applications of these viscoelastic materials comprising their use in dental applications [67], paints [68], surgical sealants, in tire technology for reduced rolling resistance [69], for packaging and acoustic insulations, even in oil and gas industry [70] too. Beyond such properties and useful applications these materials are also environmentally friendly which motivates for their widespread use.

8. Conclusions

In conclusion, our exploration of viscoelasticity has covered the fundamental principles of rheology and the nuanced behaviors exhibited by materials under dynamic conditions. Key metrics such as the Deborah number, storage modulus, and loss modulus have been discussed, providing insights into the exciting responses of materials. Also, when we look at specific applications, we can see how viscoelasticity affects a variety of industries, including engineering and healthcare. We have interpreted the complexity of material's behavior under various settings by navigating constitutive models. Along the way, we encountered experimental methods that gave us the means to discover the mysteries buried in materials. Our analysis of constitutive models, which included linear, non-linear, and quasi-linear viscoelasticity, clarified the intricate relationship between viscosity and elasticity that controls material behavior. The dynamic character of materials, as demonstrated by tests for creep, stress relaxation, and dynamic behavior, emphasizes how crucial it is to understand their behavior for real-world applications. A complete picture is painted by the progression from grasping the fundamentals to investigating sophisticated models and experimental methods.

Acknowledgements

Author would like to thank Central University of Himachal Pradesh, Shahpur Kangra (H.P.) and University Grant Commission for providing the research facilities and financial support.

Conflict of interest

The authors declare no conflict of interest.

Author details

Rahul Sharma and Noor Jahan*
Central University of Himachal Pradesh, Kangra, India

*Address all correspondence to: noor.jahan@hpcu.ac.in

References

[1] Chakraborty S. Dynamics and stability of a non-Newtonian falling film. Mechanics [physics.med-ph]. Université Pierre et Marie Curie - Paris VI; 2012. English. (NNT: 2012PAO66368). (tel-00828305)

[2] Liu G, Gao F, Wang D, Liao WH. Medical applications of magnetorheological fluid: A systematic review. Smart Materials and Structures. 2022;**31**(4):043002

[3] Kumar S, Sehgal R, Wani MF, Sharma MD, Ziyamukhamedova U, Dar TA. Next-generation ecofriendly MR fluid: Hybrid GO/ 3 encapsulated carbonyl iron microparticles with improved magnetorheological, tribological, and corrosion resistance properties. Carbon. 2023;**214**: 118331

[4] González-Aviña JV, Hosseinpoor M, Yahia A, Durán-Herrera A. New biopolymers as viscosity-modifying admixtures to improve the rheological properties of cement-based materials. Cement and Concrete Composites. 2024; **146**:105409

[5] Enhanced Rheological Properties and Performance of Viscoelastic Surfactant Fluids with Embedded Nanoparticles – ScienceDirect [Internet]. Available from: https://www.sciencedirect.com/science/article/abs/pii/S1359029418301559 [Accessed: February 2, 2024]

[6] Puljiz M, Menzel AM. Memory-based mediated interactions between rigid particulate inclusions in viscoelastic environments. Physical Review E. 2019; **99**(1):012601

[7] Monsia MD. A simplified nonlinear generalized Maxwell model for predicting the time dependent behavior of viscoelastic materials. World Journal of Mechanics. 2011;**01**(03):158-167

[8] Nakra BC. Vibration control in machines and structures using viscoelastic damping. Journal of Sound and Vibration. 1998;**211**(3): 449-466

[9] Roy N, Saha N, Kitano T, Saha P, Zatloukal M. Importance of Viscoelastic property measurement of a new hydrogel for health care. In: AIP Conference Proceedings [Internet]. Zlin (Czech Republic): AIP; 2009. pp. 210-216. Available from: https://pubs.aip.org/aip/acp/article/1152/1/210-216/864651

[10] Desbois L, Tchoreloff P, Mazel V. Characterization and modeling of the viscoelasticity of pharmaceutical tablets. International Journal of Pharmaceutics. 2020;**587**:119695

[11] Fancher IS, Levitan I. Endothelial inwardly-rectifying K+ channels as a key component of shear stress-induced mechanotransduction. In: Current Topics in Membranes [Internet]. Amsterdam, Netherlands: Elsevier, Academic Press; 2020. pp. 59-88. Available from: https://linkinghub.elsevier.com/retrieve/pii/S1063582320300119 [Accessed: January 22, 2024]

[12] Sasaki N. Viscoelastic properties of biological materials. In: De Vicente J, editor. Viscoelasticity – From Theory to Biological Applications [Internet]. London, UK: IntechOpen; 2012. Available from: http://www.intechopen.com/books/viscoelasticity-from-theory-to-biological-applications/viscoelastic-properties-of-biological-materials

[13] Saini AK, Radu T, Paritosh K, Kumar V, Pareek N, Tripathi D, et al.

Bioengineered bioreactors: A review on enhancing biomethane and biohydrogen production by CFD modeling. Bioengineered. 2021;**12**(1): 6418-6433

[14] Zhang Y, Yu G, Yu L, Siddhu MAH, Gao M, Abdeltawab AA, et al. Computational fluid dynamics study on mixing mode and power consumption in anaerobic mono- and co-digestion. Bioresource Technology. 2016;**203**: 166-172

[15] Toledo RT, Singh RK, Kong F. Flow of fluids. In: Fundamentals of Food Process Engineering [Internet]. Cham: Springer International Publishing; 2018. pp. 81-133. Available from: http://link.springer.com/10.1007/978-3-319-90098-8_5

[16] Bingham EC. An Investigation of the Laws of Plastic Flow, [Internet]. Amazon. Govt. Print. Off; 1917. Available from: https://www.amazon.com/investigation-plastic-Bulletin-Bureau-Standards/dp/B0008ANKFO [Accessed: March 26, 2024]

[17] Lipscomb GG, Denn MM. Flow of Bingham fluids in complex geometries. Journal of Non-Newton Fluid Mechanics. 1984;**14**:337-346

[18] Bingham EC. Fluidity and Plasticity. New York: McGraw-Hill; 1922

[19] Alexandrou AN, Le Menn P, Georgiou G, Entov V. Flow instabilities of Herschel–Bulkley fluids. Journal of Non-Newton Fluid Mechanics. 2003; **116**(1):19-32

[20] Burgos GR, Alexandrou AN, Entov V. On the determination of yield surfaces in Herschel–Bulkley fluids. Journal of Rheology. 1999;**43**(3): 463-483

[21] Non-Newtonian fluid. In: Wikipedia [Internet]. 2024. Available from: https://en.wikipedia.org/w/index.php?title=Non-Newtonian_fluid&oldid=1196673698 [Accessed: January 23, 2024]

[22] Schramm G. A Practical Approach to Rheology and Rheometry. Karlsruhe, Federal Republic of Germany: Gebrueder Haake GmbH

[23] Picout DR, Ross-Murphy SB. Rheology of biopolymer solutions and gels. Scientific World Journal. 2003;**3**: 105-121

[24] Messaoudi SA. General stability in viscoelasticity. In: El-Amin MF, editor. Viscoelastic and Viscoplastic Materials [Internet]. London, UK: IntechOpen; 2016. Available from: http://www.intechopen.com/books/viscoelastic-and-viscoplastic-materials/general-stability-in-viscoelasticity

[25] Phan-Thien N, Mai-Duy N. Understanding Viscoelasticity: An Introduction to Rheology [Internet]. Cham: Springer International Publishing; 2017. (Graduate Texts in Physics). Available from: http://link.springer.com/10.1007/978-3-319-62000-8 [Accessed: January 4, 2024]

[26] Ycb F. Stress-strain-history relations of soft tissues in simple elongation. Biomechanics. 1972;**1972**: 181-208

[27] Vittitow MP. Viscoelasticity in Biomechanics and Aero-space Engineering. Sandia National Laboratories DOE. Available from: https://www.osti.gov/servlets/purl/1428168

[28] Hanser Books [Internet]. Polymer Rheology. Available from: https://www.hanser-elibrary.com/doi/book/

10.3139/9781569905234 [Accessed: January 21, 2024]

[29] Thompson RL, Oishi CM. Reynolds and Weissenberg numbers in viscoelastic flows. Journal of Non-Newton Fluid Mechanics. 2021;**292**:104550

[30] Poole R. The deborah and weissenberg numbers. Rheology Bulletin. 2012;**53**(2):32-39

[31] Meyers MA, Chawla KK. Mechanical Behavior of Materials. 2nd ed. Cambridge, New York: Cambridge University Press; 2009. 856 p

[32] Zhou XQ, Yu DY, Shao XY, Zhang SQ, Wang S. Research and applications of viscoelastic vibration damping materials: A review. Composite Structures. 2016;**136**:460-480

[33] El-Amin MF. Viscoelastic and Viscoplastic Materials [Internet]. 2016. Available from: https://www.intechopen.com/books/5302 [Accessed: March 26, 2024]

[34] McCrum NG, Buckley CP, Bucknall CB. Principles of Polymer Engineering. 1: 2. ed., repr. Oxford: Oxford University Press; 2010. 447 p

[35] Endc. A. First report on viscosity and plasticity. Nature. 1935;**136**(3444): 697-699

[36] Aya M, Isayev AI. Viscoelasticity. In: Rheology Concepts, Methods, and Applications [Internet]. Toronto, Ontario, Canada: Elsevier, ChemTec Publishing; 2012. pp. 43-126. Available from: https://linkinghub.elsevier.com/retrieve/pii/B9781895198492500074 [Accessed: January 21, 2024]

[37] Burgers material. In: Wikipedia [Internet]. 2023. Available from: https://en.wikipedia.org/w/index.php?title=Burgers_material&oldid=1140738164 [Accessed: January 25, 2024]

[38] Bird RB, Armstrong RC, Hassager O. Dynamics of Polymeric Liquids, Volume 1: Fluid Mechanics. 2nd ed. New York: Wiley; 1987. 649 p.

[39] On the formulation of rheological equations of state. Proceedings of the Royal Society in London Series Maths and Physical Science. 1950;**200**(1063): 523-541

[40] Macosko CW. Rheology: Principles, Measurements, and Applications. New York: VCH; 1994 550 p. (Advances in interfacial engineering series).

[41] Bernstein B, Kearsley EA, Zapas LJ. A study of stress relaxation with finite strain. Transactions. Society of Rheology. 1963;7(1):391-410

[42] Kaye A. Non-Newtonian Flow in Incompressible Fluids. Coll Aeronaut Note 134 Amp 149 [Internet]. 1962. Available from: https://repository.tudelft.nl/islandora/object/uuid%3A8097943d-264b-41cd-8176-a0d22ce19984 [Accessed: March 26, 2024]

[43] Wagner M. Analysis of time-dependent non-linear stress-growth data for shear and elongational flow of low-density branched polyethylene melt. Rheologica Acta. 1976;**15**:136-142

[44] Wagner M. Prediction of primary normal stress difference from shear viscosity data using a single integral constitutive equation. Rheologica Acta. 1977;**16**:43-50

[45] Solid Mechanics Part I [Internet]. Available from: https://pkel015.connect.amazon.auckland.ac.nz/SolidMechanicsBooks/Part_I/index.html [Accessed: January 21, 2024]

[46] Drug Delivery [Internet]. Available from: https://www.sigmaaldrich.com/IN/en/applications/materials-science-and-engineering/drug-delivery [Accessed: January 25, 2024]

[47] Garments Merchandising [Internet]. 2016. Stretch Fabric | Types of Stretch Fabric | Use of Stretch Fabric. Available from: https://garmentsmerchandising.com/stretch-fabric-definition-types/ [Accessed: January 25, 2024]

[48] Human Technology Prosthetics and Orthotics [Internet]. Prosthetic Products | Best Artificial Arm | Artificial Limbs Manufacturers. Available from: https://humantechpando.com/prosthetic-products-lower/ [Accessed: January 25, 2024]

[49] Rade DA, Deü JF, Castello DA, De Lima AMG, Rouleau L. Passive vibration control using viscoelastic materials. In: Jauregui JC, editor. Nonlinear Structural Dynamics and Damping. Cham: Springer International Publishing; 2019. pp. 119-168. (Mechanisms and Machine Science; vol. 69). Available from: http://link.springer.com/10.1007/978-3-030-13317-7_5

[50] Thurston GB, Martin A. Rheology of pharmaceutical systems: Oscillatory and steady shear of non-Newtonian viscoelastic liquids. Journal of Pharmaceutical Sciences. 1978;**67**(11):1499-1506

[51] Budai L, Budai M, Fülöpné Pápay ZE, Vilimi Z, Antal I. Rheological considerations of pharmaceutical formulations: Focus on viscoelasticity. Gels. 2023;**9**(6):469

[52] Papakonstantinou E, Roth M, Karakiulakis G. Hyaluronic acid: A key molecule in skin aging. Dermatoendocrinology. 2012;**4**(3):253-258

[53] Bickerton S, Buntain MJ, Somashekar AA. The viscoelastic compression behavior of liquid composite molding preforms. Composition Part Applied Science Manufacturing. 2003;**34**(5):431-444

[54] Dayiary M, Najar SS, Shamsi M. An experimental verification of cut-pile carpet compression behavior. Journal of the Textile Institute. 2010;**101**(6):488-494

[55] Sajjadi A, Hosseini SA, Ajeli S, Mashayekhi M. An experimental and numerical investigation of energy absorption of sportswear in seam area. Fashion Textiles. 2023;**10**(1):9

[56] Tyler DJ, Venkatraman PD. Impact resistant materials and design principles for sportswear. In: Proceedings of the 88th Textile Institute World Conference: Bridging Innovation, Research and Enterprise [Internet]. The Textile Institute; 2012. Available from: https://e-space.mmu.ac.uk/597117/ [Accessed: March 26, 2024]

[57] Azhdari E, Ferreira JA, De Oliveira P, Da Silva PM. Diffusion, viscoelasticity and erosion: Analytical study and medical applications. Journal of Computational and Applied Mathematics. 2015;**275**:489-501

[58] Liao YH, Jones SA, Forbes B, Martin GP, Brown MB. Hyaluronan: Pharmaceutical characterization and drug delivery. Drug Delivery. 2005;**12**(6):327-342

[59] Ramachandran S, Yu Y. Peptide-based Viscoelastic matrices for drug delivery and tissue repair. BioDrugs Clinical Immunotherapy Biopharmacy Gene Therapy. 2006;**20**:263-269

[60] Wang X, Jiang M, Zhou Z, Gou J, Hui D. 3D printing of polymer matrix

composites: A review and prospective. Composites. Part B, Engineering. 2017; **110**:442-458

[61] Affatato S, Ruggiero A, Merola M. Advanced biomaterials in hip joint arthroplasty. A review on polymer and ceramics composites as alternative bearings. Composites. Part B, Engineering. 2015;**83**:276-283

[62] Singh R, Pathak S, Jain K, Noorjahan KSK. Correlating the dipolar interactions induced magneto-viscoelasticity and thermal conductivity enhancements in Nanomagnetic fluids. Small. 2023;**19**(39):2205741

[63] Jahan N, Pathak S, Jain K, Pant RP. Enchancment in viscoelastic properties of flake-shaped iron based magnetorheological fluid using ferrofluid. Colloids Surface Physicochemical Engineering Aspect. 2017;**529**:88-94

[64] Oh JS, Choi SB. State of the art of medical devices featuring smart electro-rheological and magneto-rheological fluids. Journal of King Saudi University – Science. 2017;**29**(4):390-400

[65] Noorjahan, Basheed GA, Jain K, Pathak S, Pant RP. Dipolar interaction and magneto-viscoelasticity in nanomagnetic fluid. Journal of Nanoscience and Nanotechnology (California, USA: American Scientific Publishers). 1 Apr 2018;**18**(4):2746-2751

[66] Lane SS, Lindstrom RL. Viscoelastic agents: Formulation, clinical applications, and complications. Seminars in Ophthalmology. 1992;7(4): 253-260

[67] Papadogiannis D, Lakes RS, Papadogiannis Y, Tolidis K. Mechanical viscoelastic behavior of dental adhesives. Dental Materials. 2013;**29**(6):693-701

[68] Ueda K, Kanai H, Amari T. Viscoelastic properties of paint films and formability in deep drawing of pre-painted steel sheets. Progress in Organic Coating. 2002;**45**(1):15-21

[69] Zéhil GP, Gavin HP. Rolling resistance of a rigid sphere with viscoelastic coatings. International Journal of Solids and Structures. 1 Feb 2014;**51**(3):822-838

[70] Chen C, Wang S, Harwell JH, Shiau BJ. Polymer-free viscoelastic fluid for improved oil recovery. Fuel. 2021; **292**:120331

Chapter 3

Modeling, Analysis, and Numerical Solution of a Viscoelastic Contact Problem with Normal Compliance in the Context of Locking Materials

Mustapha Bouallala

Abstract

This article delves into investigating a novel contact problem involving viscoelastic materials that exhibit ideal locking behavior when in contact with a rigid foundation, exploring their unique characteristics and implications. The contact is represented using the Signorini condition, while the friction is characterized by the nonlocal Coulomb's law. We introduce the mathematical model for the viscoelastic process, derive its variational formulation, and establish the existence and uniqueness of the solution. We introduce fully discrete finite element schemes for the variational problem and derive error estimates for the approximate solution. In conclusion, we provide an example of a viscoelastic material exhibiting blocking behavior.

Keywords: locking material, Signorini conditions, Coulomb's friction, variational inequality, weak solution, finite element method, error estimate

1. Introduction

When an obstacle prevents the deformation of a material or restricts it beyond a certain threshold, the material becomes impervious to further deformation, regardless of the applied force. This results in stress changes without corresponding deformations, and this phenomenon is referred to as material locking.

The initial investigations into variational problems related to locking materials, pioneered by Prager, are referenced in [1–3] and subsequently expanded upon by Demengel and Suquet in [4, 5]. In a recent publication [6], an investigation has been presented on a Signorini-type contact problem with friction. The model incorporates a nonmonotone multivalued subdifferential condition that is contingent on the slip. Complementary to this research, the article [7] focuses on a numerical study, providing convergence results, error estimation, and numerical simulations. The publication by Sofonea [8] introduces a mathematical model that characterizes a resilient locking material featuring memory properties. Reference [9] provides insights into the mathematical and numerical examination of a static model applied to an electro-elastic locking material. The analysis incorporates considerations of contact, friction, and interactions with a conductive foundation.

The references [7, 10–13] address elliptic variational-hemivariational inequalities related to the displacement field in various contact problems with friction. These problems span across elastic, viscoelastic, and electro-elastic-visco-plastic ideally locking materials. The authors in [14] addressed a static process involving a unilateral contact problem with nonlocal Coulomb friction between a locking material and a rigid foundation. The penalty method was employed in their approach.

A notable source for a recent model explicates the quasi-static progression of unilateral contact and friction between a thermo-electro-viscoelastic body and a conductive foundation as outlined in [15–17]. In a recent investigation, Bouallala et al. [18, 19] examined the feasibility of two innovative mathematical models. The initial model tackles a frictionless contact scenario within the realm of viscoelasticity, while the subsequent model centers on a contact problem involving friction. This second model explores the interaction between a body displaying thermo-viscoelastic behavior and a foundation with thermal conductivity. In their study, the researchers employed the Kelvin-Voigt constitutive law and incorporated the Fourier law alongside a fractional time derivative.

In this current paper, we undertake a mathematical and numerical analysis of a novel quasistatic contact problem involving a viscoelastic ideally locking material and a rigid foundation. The contact is characterized by the Signorini condition, and Coulomb's law is employed to model friction.

We establish the existence and uniqueness of a weak solution through the utilization of quasi-variational inequalities and the Banach fixed-point theorem. The discrete problem is formulated using the finite element method and a backward Euler finite difference. Additionally, we demonstrate the convergence of the solution to the discrete problem.

The rest of the paper follows this structure: in Section 2, we outline the frictional contact viscoelastic with locking model, presenting the material's behavior through a nonlinear viscoelastic-locking constitutive law, equilibrium equation, and boundary conditions. In Section 3, we introduce the notations and assumptions regarding the problem's data, and we derive its variational formulation. Section 4 is dedicated to presenting the proof of weak solvability. Moving on to Section 5, we introduce a fully discrete scheme, derive error estimates, and establish convergence results. Finally, in the last section, we provide a concrete example involving a viscoelastic blocking material.

2. Contact problem involving locking material

Let $\mathbb{S}^d$ denote the space of second-order symmetric tensors on $\mathbb{R}^d$. The symbols "·"and ·‖ will denote the inner product and the Euclidean norm on $\mathbb{R}^d$ and $\mathbb{S}^d$, respectively. Specifically, for any u and v in $\mathbb{R}^d$ and for any σ and τ in $\mathbb{S}^d$, the expressions are

$$u \cdot v = u_i \cdot v_i, \quad \|v\| = (v \cdot v)^{\frac{1}{2}} \quad \text{and} \quad \sigma \cdot \tau = \sigma_{ij} . \tau_{ij}, \quad \|\tau\| = (\tau \cdot \tau)^{\frac{1}{2}}. \tag{1}$$

Here, the indices i and j range from 1 to d, and unless explicitly mentioned otherwise, we apply the summation convention to repeated indices.

We examine a rigid body initially located within an open, bounded domain $\Omega \subset \mathbb{R}^d$, where $d = 1,2,3$. The boundary Γ of the domain Ω is presumed to be Lipschitz, and it is partitioned into three distinct, measurable components: Γ_D, Γ_N, and Γ_C, with the

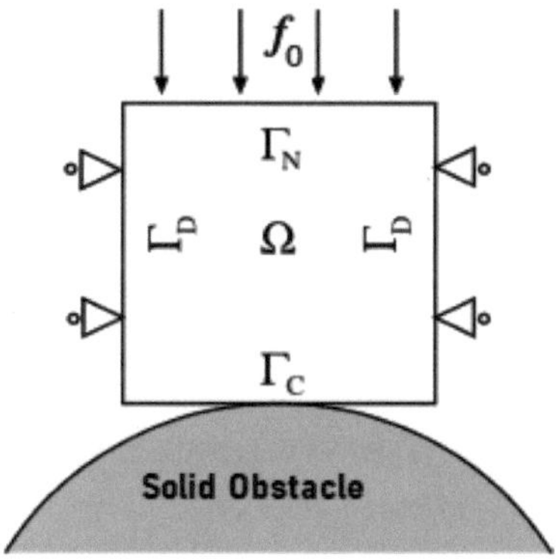

Figure 1.
Domain in the initial configuration.

condition that $mes(\Gamma_D) > 0$. In this context, let $T > 0$, and consider the time interval of interest as $[0, T]$. The body is immobilized within $\Gamma_D \times (0, T)$, causing the displacement field to be zero in that region. Meanwhile, a volume force with a density of f_0 operates in $\Omega \times (0, T)$, and surface traction with a density of f_1 applies to $\Gamma_N \times (0, T)$. At $\Gamma_C \times (0, T)$, the body has the potential to interact with a solid obstacle referred to as the foundation (**Figure 1**).

Moreover, a comma signifies partial differentiation concerning the spatial variable x, as in $u_{i,j} = \frac{\partial u_i}{\partial x_j}$. Additionally, ε and Div represent the linearized strain tensor and the deformation and the divergence operator, respectively, namely,

$$\varepsilon(u) = \left(\varepsilon_{ij}(u)\right), \quad \varepsilon_{ij}(u) = \frac{1}{2}\left(u_{i.j} + u_{j.i}\right), \quad Div\ \sigma = \left(\sigma_{ij,j}\right). \tag{2}$$

Furthermore, $\nu = (\nu_i)$ represents the outward unit normal vector at Γ. The terms $u_\nu = u.\nu$ and $u_\tau = u - u_\nu \nu$ correspond to the normal and tangential components of the vector field u on the boundary Γ.

Lastly, the normal stress $\sigma_\nu = (\sigma\nu).\nu$ and tangential stress $\sigma_\tau = \sigma\nu - \sigma_\nu \nu$ are defined on the boundary Γ.

With these considerations, we are now ready to present the formulation of the contact problem under the aforementioned assumptions.

Problem (P): Find a displacement field $u : \Omega \times (0, T) \to \mathbb{R}^d$ and a stress tensor $\sigma : \Omega \times (0, T) \to \mathbb{S}^d$, a.e. $t \in [0, T]$ such that

$$\sigma(t) \in \mathcal{V}\varepsilon(\dot{u}(t)) + \mathcal{E}\varepsilon(u(t)) + \partial(I_A(\varepsilon(u(t)))) \qquad \text{in} \quad \Omega \times (0, T), \tag{3}$$

$$Div(\sigma(u(t))) + f_0(t) = 0 \qquad \text{in} \quad \Omega \times (0, T), \tag{4}$$

$$u(t) = 0 \qquad \text{on} \quad \Gamma_D \times (0, T), \tag{5}$$

$$\sigma(t)\nu = f_1(t) \qquad \text{on} \quad \Gamma_N \times (0, T), \tag{6}$$

$$\sigma_\nu(t) \le 0, \quad u_\nu(t) - g \le 0, \quad \sigma_\nu(t)(u_\nu(t) - g) = 0 \qquad \text{on} \quad \Gamma_C \times (0, T), \tag{7}$$

$$\left.\begin{array}{l} \|\sigma_\tau\| \le \mu(\|u_\tau\|)|R\sigma_\nu(u)|, \\ \|\sigma_\tau\| < \mu(\|u_\tau\|)|R\sigma_\nu(u)| \Rightarrow \dot{u}_\tau = 0, \\ \|\sigma_\tau\| = \mu(\|u_\tau\|)|R\sigma_\nu(u)| \Rightarrow \exists \lambda \neq 0 \mid \sigma_\tau = -\lambda \dot{u}_\tau \end{array}\right\} \qquad \text{on} \quad \Gamma_C \times (0, T), \tag{8}$$

with I_A is the indicator function defined by

$$I_A(\alpha) = \begin{cases} 0, & \text{if } \alpha \in A, \\ +\infty, & \text{otherwise} \end{cases} \tag{9}$$

of a closed and convex set A given by $A = \left\{\beta \in \mathbb{S}^d, \ \|\beta\| \leq M\right\}$.

It is worth noting that Eq. (3) characterizes the constitutive behavior of viscoelastic materials subject to locking constraints, where $\mathcal{E} = (e_{ijkl})$ and $\mathcal{V} = (v_{ijkl})$ are, respectively, the elasticity tensor and the viscosity tensor. On the other hand, relation (4) delineates the equilibrium condition for the stress-displacement field. Conditions (5) and (6) describe the displacement and traction boundary conditions, respectively. Additionally, relation (7) corresponds to the normal compliance contact condition on Γ_C. Finally, conditions (8) embody Coulomb's friction law, where μ denotes the coefficient of friction and R serves as a regularization operator; for more details, see [20, 21].

3. Variational formulation

To introduce the weak formulation of **Problem (P)**, we require additional notation and preliminary concepts. Let X denote a Banach space, and let T be a positive real number.

We employ the conventional notation for the spaces $L^p(0,T;X)$ and $W^{k,p}(0,T;X)$, where $1 \leq p \leq \infty$ and $k = 1, \ldots,$ equipped with the corresponding norms.

$$\|u\|^p_{L^p(0,T;X)} = \int_0^T \|u\|^p_X dt. \tag{10}$$

We also make use of Hilbert spaces.

$$\begin{aligned} H(\Omega) &= L^2(\Omega)^d = \{u = (u_i), \ u_i \in L^2(\Omega)\}, \\ H_1(\Omega) &= H^1(\Omega)^d, \\ \mathcal{H} &= \{\tau = (\tau_{ij}) \mid \tau_{ij} = \tau_{ji} \in L^2(\Omega)\}. \end{aligned} \tag{11}$$

These spaces are real Hilbert spaces equipped with inner products.

$$\begin{aligned} &(u,v)_{H(\Omega)} = \int_\Omega u_i v_i dx, \ (\sigma,\tau)_{\mathcal{H}} = \int_\Omega \sigma_{ij}\tau_{ij} dx, \\ &(u,v)_{H_1(\Omega)} = (u,v)_{H(\Omega)} + (\varepsilon(u),\varepsilon(v))_{\mathcal{H}}. \end{aligned} \tag{12}$$

Considering the boundary condition (5), we define a closed subspace of $H_1(\Omega)$ as follows:

$$V = \{v \in H_1(\Omega) \ : \ v = 0 \ \text{on} \ \Gamma_D\}. \tag{13}$$

The set of admissible displacements

$$K_1 = \{v \in V \ / \ v_\nu - g \leq 0 \ \text{on} \ \Gamma_C\}, \tag{14}$$

and the closed convex

$$K_2 = \{v \in V \ / \ ; |\varepsilon(v)| \leq M \ \text{ a.e.on } \ \Omega\}. \tag{15}$$

We define V^* as the dual space of V and by identifying $L^2(\Omega)^d$ with its own dual. Also, it holds $V \subset L^2(\Omega)^d = \left(L^2(\Omega)^d\right)^* \subset V^*$. The duality pairing between V^* and V is denoted as $(\cdot,\cdot)_{V^* \times V}$, and the norm in V^* is represented by $\|\cdot\|_{V^*}$.

For a given mapping φ defined on the space X, where X^* represents its topological space, $\partial\varphi$ signifies the subdifferential of φ as defined in (see [22]):

$$\partial\varphi(x) = \{z \in X^* \ | \ \langle z, y - x \rangle \leq \varphi(y) - \varphi(x), \ \ \forall y \in X\}. \tag{16}$$

Given that $meas(\Gamma_D) > 0$ and Korn's inequality is satisfied (refer to [23]),

$$\|\varepsilon(v)\|_{\mathcal{H}} \geq c_t \|v\|_{H_1(\Omega)}, \ \ \forall v \in V, \tag{17}$$

where c_t is a nonnegative constant determined solely by the properties of Ω and Γ_D.

This inequality plays a crucial role in the theory of linear elasticity, where the symmetric part of the gradient serves as a measure of the deformation experienced by an elastic body when deformed by a given vector function. Thus, this inequality is an essential tool as an a priori estimate in the context of linear elasticity theory. For more details, refer to [24].

In the vector space V, we contemplate the inner product defined by

$$(u, v)_V = (\varepsilon(u), \varepsilon(v))_{\mathcal{H}}, \ \ \|u\|_V^2 = (u, u), \tag{18}$$

and the norm associated with it, denoted by $\|v\|_V$, is equivalent on V to the conventional norm $\|\varepsilon(v)\|_{\mathcal{H}}$ in $\|\cdot\|_{H_1(\Omega)}$.

Furthermore, leveraging the Sobolev trace theorem, there exists a positive constant c_0 dependent solely on Ω, Γ_C, and Γ_D such that

$$\|v\|_{L^2(\Gamma_C)^d} \leq c_0 \|v\|_V, \ \ \forall v \in V. \tag{19}$$

To proceed with the investigation of the mechanical **Problem (P)**, we require the following assumptions:

(HP_1) The elasticity operator $\mathcal{E} : \Omega \times \mathbb{S}^d \to \mathbb{S}^d$ satisfies the usual properties of symmetry, boundedness, ellipticity, and continuity

i. $f_{ijkl} = f_{jikl} = f_{lkij} \in L^\infty(\Omega)$,

ii. $f_{ijkl}(x)\xi_k\xi_l \geq m_a\|\xi\|^2$, for all ξ in $\mathbb{S}^d$, and for all x in Ω, with $m_a > 0$,

iii. $\|(\mathcal{E}\varepsilon(u), \varepsilon(v))_{\mathcal{H}}\| \leq M_a\|u\|_V\|v\|_V$.

(HP_2) The viscosity tensor $\mathcal{V} : \Omega \times \mathbb{S}^d \to \mathbb{S}^d$ satisfy

i. $v_{ijkl} = v_{jikl} = v_{lkij} \in L^\infty(\Omega)$,

ii. $v_{ijkl}(x)\xi_k\xi_l \geq m_b\|\xi\|^2$, for all ξ in $\mathbb{S}^d$, for all x in Ω, with $m_b > 0$,

iii. $\|(\mathcal{V}\varepsilon(u), \varepsilon(v))_{\mathcal{H}}\| \leq M_b\|u\|_V\|v\|_V$.

The symmetry of the elasticity and viscosity tensors implies that the order of force application does not affect the result. Moreover, this often corresponds to conservation laws.

The continuity of tensors ensures that small changes in input lead to small changes in output. In the context of elasticity and viscosity, it is crucial for the harmonious behavior of materials under deformation or flow. Physically, continuous operators reflect the concept that a material's response to gradual changes in conditions is itself gradual. Sudden changes in stress or deformation are avoided.

Coercivity is a property that ensures the operator "pushes back" significant inputs, preventing unbounded behavior. In elasticity and viscosity, coercivity is often associated with the stability and well-posedness of mathematical models. In mechanics, coercivity ensures that materials resist extreme deformations or rapid changes in stress, thereby contributing to the physical stability of the system.

(HP_3) The forces, the traction, the initial conditions, and the gap function satisfy

$$\begin{aligned} &f_0 \in W^{1,\infty}\left(0,T;L^2(\Omega)^d\right), \quad f_1 \in W^{1,\infty}\left(0,T;L^2(\Gamma_N)^d\right), \\ &u_0 \in V, \quad g \in L^2(\Gamma_C), \quad g \geq 0. \end{aligned} \tag{20}$$

The regularities of these elements not only are important mathematically for the well-posedness of problems but also have direct physical implications, contributing to the stability, reliability, and accuracy of models in various scientific and engineering applications.

(HP_4) The coefficient of friction $\mu : \Gamma_C \times \mathbb{R}^+ \to \mathbb{R}^+$ satisfies the following:

i. There exists $L_\mu > 0$, for all a, b in $\mathbb{R}^+$, $|\mu(.,a) - \mu(.,b)| < L_\mu|a - b|$ a.e. on Γ_C.

ii. For all a in $\mathbb{R}^+$, the mapping $x \mapsto \mu(x,a)$ is measurable on Γ_C.

iii. For all a in $\mathbb{R}^+$, the mapping $x \mapsto \mu(x,a)$ is μ^* -bounded, a.e. on Γ_C, with

$$\mu^* = \sup_{t \in [0,T]} \|\mu\|_{L^\infty(\Gamma_C)}. \tag{21}$$

iv. The mapping $R : H'_{\Gamma_C} \to L^\infty(\Gamma_C)$ is linear compact and continuous with $c_R = \|R\|$.

This implies that the friction coefficient is defined at each point on the contact boundary, and its value depends on the magnitude of the applied normal forces, limited to positive real numbers, reflecting the physical nature of the friction coefficient as a nonnegative quantity. The friction coefficient may vary in space, reflecting the heterogeneity of friction properties in different contact regions. This situation is often encountered in real-world scenarios where surfaces may have different textures or materials. The friction coefficient is sensitive to changes in the applied normal force at the contact interface. This reflects the physical reality that the friction behavior between surfaces can be influenced by the magnitude of compression forces.

Subsequently, employing Riesz's representation theorem, we characterize the elements $f \in V$ as follows:

$$(f(t), v)_{V^* \times V} = \int_{\Omega} f_0(t) \cdot v dx + \int_{\Gamma_N} f_1(t) v da. \tag{22}$$

We represent the bilinear and symmetric mappings $a : V \times V \to \mathbb{R}$ and $c : V \times V \to \mathbb{R}$ as follows:

$$a(u(t), v) =: (\mathcal{E}\varepsilon(u(t)), \varepsilon(v))_{\mathcal{H}} \text{ and } \quad b(u(t), v) =: (\mathcal{V}\varepsilon(u(t)), \varepsilon(v))_{\mathcal{H}}. \tag{23}$$

We establish the mappings $j : V \times V \to \mathbb{R}$ as follows:

$$j(u(t), v) = \int_{\Gamma_C} \mu(\|u_\tau(t)\|) |R\sigma_\nu(u(t))| \|v_\tau\|_V da, \quad \forall u, v \in V. \tag{24}$$

Now, employing a standard variational approach, it becomes evident that if u satisfies the conditions (3)–(6) and (22)–(24), then for almost every $t \in]0, T[$ and for all $v \in V$,

$$\begin{aligned} b(\dot{u}(t), v - \dot{u}(t)) + a(u(t), v - \dot{u}(t)) + \langle Y(\varepsilon(u(t))), \varepsilon(v) - \varepsilon(\dot{u}(t)) \rangle \\ + j(u(t), v) - j(u(t), \dot{u}(t)) \geq (f(t), v - \dot{u}(t))_{V^* \times V}, \end{aligned} \tag{25}$$

with $Y(\varepsilon(u(t))) \in \partial I_{K_2}(\varepsilon(u(t)))$.
Exploiting the convex nature of I_A yields:

$$\langle Y(\varepsilon(u(t)), \varepsilon(u(t) - \dot{u}(t))) \rangle \leq I_{K_2}(\varepsilon(v)) - I_{K_2}(\varepsilon(\dot{u}(t))) = 0, \quad \forall v \in K_2. \tag{26}$$

By incorporating all these assumptions, we derive the weak formulation as a variational inequality.

Problem (PV): Find a displacement field $u :]0; T[\to \mathbb{R}^d$ a.e. $t \in]0; T[$ and for all $v \in K_1 \cap K_2$ such that:

$$u \in K_1 \cap K_2, \tag{27}$$

$$\begin{aligned} c(\dot{u}(t), v - \dot{u}(t)) + a(u(t), v - \dot{u}(t)) + j(u(t), v) - j(u(t), \dot{u}(t)) \\ \geq (f(t), v - \dot{u}(t))_{V^* \times V}. \end{aligned} \tag{28}$$

4. Existence and uniqueness of the solution

Theorem 4.1. *Under the assumptions* $(HP_1) - (HP_4)$, *and the following condition*

$$m_b > \left(\mu^* c_0^2 + c_R L_\mu c_0^2\right) \tag{29}$$

a unique solution to **Problem (PV)** exists, satisfying the following regularity:

$$u \in W^{2,\infty}(0, T; V). \tag{30}$$

Proof. Under the assumptions (HP_2), the viscosity operator is bilinear, symmetric, continuous, and coercive.

Similar to (HP_1), the operator a is bilinear, symmetric, and continuous.
By (HP_4), the operator j is a proper convex l.s.c.
Now, let u_1, u_1, v_1, and v_1 in V, using (24), we have that

$$\begin{aligned} j(u_1(t),v_2)+j(u_1(t),v_1)+j(u_2(t),v_1)+j(u_2(t),v_2) \\ = \int_{\Gamma_C} \mu(\|u_{1_\tau}\|_V)(R\sigma_\nu(u_1(t)) - R\sigma_\nu(u_2(t)))(\|u_{1_\tau}\|_V - \|u_{2_\tau}\|_V)da \\ + \int_{\Gamma_C} |R\sigma_\nu(u_2(t))|(\mu(\|u_{1_\tau}\|_V) - \mu(\|u_{2_\tau}\|_V))(\|u_{1_\tau}\|_V - \|u_{2_\tau}\|_V)da. \end{aligned} \tag{31}$$

Taking into account the continuity of R, (19), the conditions presented in (HP_4), and after calculation, we deduce that

$$\begin{aligned} j(u_1(t),v_2)+j(u_1(t),v_1)+j(u_2(t),v_1)+j(u_2(t),v_2) \\ \le \left(\mu^* c_0^2 + c_R L_\mu c_0^2\right)\|u_1(t) - u_2(t)\|_V\|v_1 - v_2\|_V. \end{aligned} \tag{32}$$

Then, considering the regularities of f_0 and f_1 as stated in (HP_3), we conclude that $f \in W^{1,\infty}(0,T;V)$.

Finally, by combining all these results with (29) and utilizing the theorem presented in Sofonea and Matei ([25], pp. 67–70), the **Problem (PV)** has a unique solution that satisfies (30).

5. Fully discrete approximation

In this section, we present a fully discrete numerical scheme for solving **Problem (P)** and derive an optimal error estimate. We employ the finite element space V^h and introduce a time interval partition $[0,T]$ as follows: $0 = t_0 < t_1 < \ldots < t_N = T$.

Consider time step sizes $k_n > 0$, defined as $k_n = t_n - t_{n-1}$ for $n = 1, \ldots, N$. We permit a nonuniform partition of the time interval and denote $k = \max_n k_n$ as the maximal step size.

For a continuous function $u(t)$ taking values in a function space, we express $u_i = u(t_i)$ for $0 \le i \le N$. We represent the difference as $\Delta w_n = w_n - w_{n-1}$ and the corresponding divided difference as $\delta w = \frac{\Delta w}{k_n}$.

Consider a regular family of triangular finite element partitions $\{T^h\}$ defined on $\overline{\Omega}$, which is compatible with the boundary decomposition $\Gamma = \overline{\Gamma}C \cup \overline{\Gamma}D \cup \overline{\Gamma}_N$.

We proceed to define a finite element space $V^h \subset V$, $K_1^h = K_1 \cap V^h$, and $K_2^h = K_2 \cap V^h$ for the approximations of the displacement field u as follows:

$$V^h = \left\{v^h \in \left[C(\overline{\Omega})\right]^d;\ v^h_{|Tr} \in [\mathbb{P}_1(Tr)]^d \ \ \forall Tr \in T^h;\ v^h = 0 \text{ on } \overline{\Gamma}_D\right\}, \tag{33}$$

The fully discrete approximation method is founded on a backward Euler scheme, taking the following form:

Problem (PVHK): Find a displacement field $\{u_n^{hk}\} \subset K_1^h \cap K_2^h$ for all $v^h \in V^h$ and $n = 1, \ldots, N$

$$\begin{aligned} b\left(w_n^{hk}, v^h - w_n^{hk}\right) + a\left(u_{n-1}^{hk}, v^h - w_n^{hk}\right) \\ + j\left(u_{n-1}^{hk}, v^h\right) - j\left(u_{n-1}^{hk}, w_n^{hk}\right) \ge \left(f_n, v^h - w_n^{hk}\right), \\ u_0^{hk} = u_0^h, \end{aligned} \tag{34}$$

with $u_0^h \in V^h$ is an approximate of u_0.

To simplify the notation further, we introduce the velocity.

$$w_n^{hk} = \delta u_n^{hk}, \ n = 1, \dots, N, \quad \text{and} \quad u_n^{hk} = \sum_{j=1}^{n} k_j w_j^{hk} + u_0^h. \tag{35}$$

Applying a discrete analogue of Theorem 4.1, we observe that given $u_{n-1}^{hk} \in V^h$, **Problem (PVHK)** possesses a unique solution $u_n^{hk} \in V^h$..

Next, we establish the following convergence result for the fully discrete solution.

Theorem 5.1. *Assume the initial values* $u_0^h \in V^h$ *and the assumptions* $(HP_1) - (HP_4)$. *Then, if*

$$\|u_0 - u_0^h\|_V \to 0, \quad as \quad h \to 0, \tag{36}$$

the fully discrete solution of the **Problem (PVHK)** converges:

$$\max_{1 \le n \le N} \left\{ \|u_n - u_n^{hk}\|_V + \|w_n - w_n^{hk}\|_V \right\} \to 0, \quad as \quad h, \ k \to 0. \tag{37}$$

Proof. Substitute $v = w_n^{hk}$ into (28) at $t = t_n$, and we obtain:

$$\begin{aligned} &b\left(w_n^{hk}, w_n^{hk} - w_n\right) + a\left(u_n, w_n^{hk} - w_n\right) \\ &j\left(u_n, w_n^{hk}\right) - j\left(u_n, w_n\right) \ge \left(f_n, w_n^{hk} - w_n\right). \end{aligned} \tag{38}$$

Summing up (34) with $v^h = v_n^h$ and (34)

$$\begin{aligned} b\left(w_n^{hk}, v_n^h - w_n^{hk}\right) + v\left(w_n, w_n^{hk} - w_n\right) + a\left(u_{n-1}^{hk}, v_n^h - w_n^{hk}\right) + a\left(u_n, w_n^{hk} - w_n\right) \\ +j\left(u_{n-1}^{hk}, v_n^h\right) - j\left(u_{n-1}^{hk}, w_n^{hk}\right) + j\left(u_n, w_n^{hk}\right) - j\left(u_n, w_n\right) \\ \ge \left(f_n, v_n^h - w_n^{hk}\right) + \left(f_n, w_n^{hk} - w_n\right). \end{aligned} \tag{39}$$

With this equality

$$\begin{aligned} b\left(w_n - w_n^{hk}, w_n - w_n^{hk}\right) = b\left(w_n^{hk}, w_n^{hk} - v_n^h\right) + b\left(w_n, w_n - v_n^h\right) \\ -b\left(w_n^{hk}, w_n - v_n^h\right) + b\left(w_n, v_n^h - w_n^{hk}\right). \end{aligned} \tag{40}$$

We deduce that

$$\begin{aligned} b\left(w_n - w_n^{hk}, w_n - w_n^{hk}\right) = R^{hk} + R_j^{hk} + b\left(w_n^{hk} - w_n, v_n^h - w_n\right) \\ +a\left(u_{n-1}^{hk} - u_n, v_n^h - w_n^{hk}\right), \end{aligned} \tag{41}$$

where

$$\begin{aligned} R^{hk} = c\left(w_n, v_n^h - w_n\right) + a\left(u_n, v_n^h - w_n\right) \\ +j\left(u_n, v_n^h\right) - j\left(u_n, w_n\right) - \left(f_n, v_n^h - w_n\right), \end{aligned} \tag{42}$$

and

$$R_j^{hk} = j\left(u_{n-1}^{hk}, v_n^h\right) - j\left(u_{n-1}^{hk}, w_n^{hk}\right) + j\left(u_n, w_n^{hk}\right) - j\left(u_n, v_n^h\right). \tag{43}$$

Under the assumptions $(HP_1) - (HP_2)$, we obtain

$$\begin{aligned} m_b\left\|w_n - w_n^{hk}\right\|^2 \leq & M_b\left\|w_n^{hk} - w_n\right\|_V\left\|v_n^h - w_n\right\|_V \\ & + M_e\left\|u_{n-1}^{hk} - u_n\right\|_V\left\|v_n^h - w_n^{hk}\right\|_V + \left|R^{hk}\right| + \left|R_j^{hk}\right|. \end{aligned} \tag{44}$$

Due to (24) and the assumption (HP_4) and applying the inequality

$$ab \leq \alpha a^2 + \frac{1}{4\alpha} b^2, \quad \forall \alpha > 0, \tag{45}$$

we found the following estimate

$$\begin{aligned} \left|R_j^{hk}\right| & = \left|j\left(u_{n-1}^{hk}, v_n^h\right) - j\left(u_{n-1}^{hk}, w_n^{hk}\right) + j\left(u_n, w_n^{hk}\right) - j\left(u_n, v_n^h\right)\right| \\ & = \left(\mu^* c_0^2 + c_R L_\mu c_0^2\right)\left\|u_{n-1}^{hk} - u_n\right\|_V\left\|w_n^{hk} - v_n^h\right\|_V. \\ & = c\left\{\left\|u_n - u_{n-1}^{hk}\right\|_V^2 + \left\|v_n^h - w_n^{hk}\right\|_V^2\right\}. \end{aligned} \tag{46}$$

Here, $\left\|v_n^h - w_n^{hk}\right\|_V$ will be bounded as follows:

$$\left\|v_n^h - w_n^{hk}\right\|_V \leq \left\|v_n^h - w_n\right\|_V + \left\|w_n - w_n^{hk}\right\|_V. \tag{47}$$

Combining (44), (46), and (47), we obtain the following relation

$$\left\|w_n^h - w_n^{hk}\right\|_V \leq c\left\{\left\|v_n^h - w_n\right\|_V + \left\|u_{n-1}^{hk} - u_n\right\|_V + \left|R^{hk}\right|^{\frac{1}{2}}\right\}. \tag{48}$$

Following the same procedures outlined in Han and Sofonea ([26], pp. 565–567), we obtain the following estimate

$$\begin{aligned} & \max_{1 \leq n \leq N} \left\{\left\|u_n - u_n^{hk}\right\|_V + \left\|w_n - w_n^{hk}\right\|_V\right\} \\ & \leq c \max_{1 \leq n \leq N} \inf_{v_n^h \in V^h} \left\{\left\|v_n^h - w_n\right\|_V + \left|R_n^{hk}\right|^{\frac{1}{2}}\right\} \\ & + \left\|u_0^h - u_0\right\|_V + c\left(I_k(w) + k\|u\|_{C^1(0,T;V)}\right). \end{aligned} \tag{49}$$

where

$$I_k(w) = \sum_{j=1}^{N} \int_{t_{j-1}}^{t_j} \left\|w_j - w_s\right\|_V ds. \tag{50}$$

We take $v_n^h = \mathcal{P}^h(w_n)$ with $\mathcal{P}^h: \ V \rightarrow V$ is a projection operator defined by

$$\left\|v - \mathcal{P}_v^h\right\|_V = \inf_{v^h \in V^h} \left\|v - v^h\right\|_V, \ \forall v \in V. \tag{51}$$

Utilizing Lemma 5.4, established in Han and Sofonea [26], we deduce the result stated in Theorem 5.1.

6. Appropriate example

In this section, we illustrate an example of a viscoelastic blocking material. It is worth noting that viscoelasticity is employed to characterize the behavior of reversible materials that are sensitive to the rate of deformation, including polymers, concrete, wood, rubber, and the like.

Example: Exploring the coupled mechanosorptive and viscoelastic behavior of wood. The mechanosorptive behavior includes the hygro-lock phenomenon, characterized by deformation blockage during the drying phase under stress.

The polymeric composition of wood, including cellulose, hemicellulose, and lignin, contributes to its viscoelastic behavior. This behavior is observed through the gradual increase in deformation over time under constant stress (creep) or the relaxation of stress over time under constant deformation. It is characterized by its time-dependent nature and is influenced by various factors, including the external atmosphere.

Assuming a partition of the deformation, we can express

$$\varepsilon(t) = \varepsilon_1(t) + \varepsilon_2(t). \tag{52}$$

Here, ε_1 represents the elastic deformation, compensating for the blocked part caused by the variation of the modulus of elasticity during drying and ε_2 denotes the viscoelastic deformation.

The dual form associated with the constraint set is expressed as

$$\sigma(t) = \sigma_1(t) + \sigma_2(t). \tag{53}$$

7. Conclusion

In this study, we have addressed a novel viscoelastic contact problem involving Coulomb friction for locking materials. The contact was modeled using the Signorini condition law. We established the existence and uniqueness of a weak solution and derived error estimates for the approximate solution. The proofs relied on arguments involving time-dependent variational inequalities, finite element methods, and Euler's forward scheme.

We emphasize that the key distinction between contact problems with materials without blocking and those involving blocking materials, as considered here, lies in the nature of the material's behavior law, which is of a differential type.

We conclude our study with the following outlook:

1. The model can be explored with alternative behavior laws such as electro-viscoelasticity, thermo-viscoelasticity, thermo-electro-viscoelasticity, and so on.

2. Investigating problems of a similar nature with diverse types of contact, friction, and locking effects could be a promising avenue for future research.

3. Addressing the same problem in the dynamic case poses greater challenges.

4. Conducting numerical simulations using finite element methods or other approaches represents a compelling direction for future research.

Conflict of interest

The authors declare no conflict of interest.

Author details

Mustapha Bouallala
Department of Mathematics and Computer Science, Polydisciplinary Faculty of Safi, Cadi Ayyad University, Marrakech, Morocco

*Address all correspondence to: bouallalamustaphaan@gmail.com

References

[1] Prager W. On ideal locking materials. Transactions of the Society of Rheology. 1957;**1**(1):169-175

[2] Prager W. Elastic solids of limited compressibility. In: Proceedings of the 9th International Congress of Applied Mechanics. Vol. 5. 1958. pp. 205-211

[3] Prager W. On elastic, perfectly locking materials. In: Applied Mechanics: Proceedings of the Eleventh International Congress of Applied Mechanics Munich (Germany) 1964. Berlin, Heidelberg: Springer; 1966. pp. 538-544

[4] Demengel F, Suquet P. On locking materials. Acta Applicandae Mathematica. 1986;**6**(2):185-211

[5] Demengel F. Displacements bounded deformation and measures stress. Annals of Superior School of Pise. 1972

[6] Barboteu M, Han W, Migórski S. On numerical approximation of a variational-hemivariational inequality modeling contact problems for locking materials. Computers & Mathematics with Applications. 2019;**77**(11): 2894-2905

[7] Migórski S, Ogorzały J. A variational-hemivariational inequality in contact problem for locking materials and nonmonotone slip dependent friction. Acta Mathematica Scientia. 2017;**37**(6): 1639-1652

[8] Sofonea M. History-dependent inequalities for contact problems with locking materials. Journal of Elasticity. 2019;**134**(2):127-148

[9] Essoufi EH, Zafrar A. Dual methods for frictional contact problem with electroelastic-locking materials. Optimization. 2021;**70**(7):1581-1608

[10] Migórski S, Ochal A. Dynamic bilateral contact problem for viscoelastic piezoelectric materials with adhesion. Nonlinear Analysis: Theory, Methods & Applications. 2008;**69**(2):495-509

[11] Migórski S, Ochal A, Sofonea M. Analysis of a piezoelectric contact problem with subdifferential boundary condition. Proceedings of the Royal Society of Edinburgh Section A: Mathematics. 2014;**144**(5):1007-1025

[12] Sofonea M. A nonsmooth static frictionless contact problem with locking materials. Analysis and Applications (World Scientific). 2018;**16**(6):851-874

[13] Sofonea M, Migorski S. Variational-hemivariational inequalities with applications. Chapman and Hall/CRC; 2017

[14] Bourichia S, El-H. Essoufi. Penalty method for an unilateral contact problem with Coulomb's friction for locking materials. International Journal of Mathematical Modelling & Computations. 2016;**6**(1):61-81

[15] Essoufi EH, Alaoui M, M. Bouallala Quasistatic thermo-electro-viscoelastic contact problem with Signorini and Tresca's friction electronic. Journal of Differential Equations. 2019;**5**:1-21

[16] Bouallala M, El-H Essoufi M. Alaoui, Numerical analysis of the penalty method for unilateral contact problem with Tresca's friction in thermo- electro-visco-elasticity. Eurasian Journal of Mathematical and Computer Applications. 2020;**8**(3):12–32

[17] Bourichi S, Essoufi E. Penalty method for unilateral contact problem with coulomb's friction for locking

material. International Journal of Mathematical Modelling & Computations. 2016;**6**(1 (WINTER)): 61-81

[18] Bouallala M, Essoufi EL-H. A thermo-viscoelastic fractional contact problem with normal compliance and Coulomb's friction. Journal of Mathematical Physics, Analysis, Geometry. 2021;**17**(3):280-294

[19] Bouallala M, Essoufi EH, Nguyen VT, Pang W. A time-fractional of a viscoelastic frictionless contact problem with normal compliance. The European Physical Journal Special Topics. 2023;**232**(14):2549-2558

[20] Martins JAC, Oden JT. Existence and uniqueness results for dynamic contact problems with nonlinear normal and friction interface laws. Nonlinear Analysis: Theory, Methods and Applications. 1987;**11**(3):407-428

[21] Oden JT, Pires EB. Nonlocal and nonlinear friction laws and variational principles for contact problems in elasticity. Journal of Applied Mechanics. 1983;**50**(1):7-76

[22] Ekeland I, Temam R. Convex analysis and variational problems. Society for Industrial and Applied Mathematics. 1999

[23] Necas J, Hlavácek I. Mathematical theory of elastic and elasto-plastic bodies: An introduction. Elsevier; 2017

[24] Horgan CO. Korn's inequalities and their applications in continuum mechanics. SIAM Review. 1995;**37**(4): 491-511

[25] Sofonea M, Matei A. Variational inequalities with applications: A study of antiplane frictional contact problems. Vol. 18. Springer Science & Business Media; 2009

[26] Han W, Sofonea M. Evolutionary variational inequalities arising in viscoelastic contact problems. SIAM Journal on Numerical Analysis. 2002; **38**(2):556-579

Section 3

Rheological Measurement in Petroleum Industry

Chapter 4

Emulsion Rheology: Applications and Measuring Techniques in Upstream Petroleum Operations

Iskandar Bin Dzulkarnain, Muhammad Mohsin Yousufi and Mysara Eissa Mohyaldinn Elhaj

Abstract

Emulsions play a prominent role in the petroleum exploration and production fields. They act as working fluids for drilling exploration wells, cementing the completion assembly, and helping to enhance hydrocarbon production. Moreover, the liquid hydrocarbons produced are often in the form of emulsions. The rheological analysis is crucial for the effectiveness of emulsion-based working fluids as well as the optimization and refinement of water-engulfed crude oil produced. The chapter covers the vast role of emulsions in the petroleum upstream section and the rheological techniques applied (such as rotational, oscillatory, extensional rheometry, and microfluidics) to better understand the flow characteristics of the non-Newtonian fluid in multiple applications.

Keywords: rheology, emulsion rheology, emulsified acids, rheometer, Pickering emulsion

1. Introduction

Emulsions are colloidal systems consisting of two immiscible liquids combined together with the help of an emulsifier [1]. These colloids are found in various industrial applications including cosmetics, pharmaceuticals, food, energy, and fuels [2]. The coexistence of generally two unmixable fluids provides unique characteristics that are not found otherwise, such as providing enhanced transportation and controlled release capabilities for various substances with required stability under varying thermal and mechanical conditions. These colloidal suspensions allow for tailor-made rheological characteristics with respect to viscosity, flow rate, sweep efficiency, phase ratio, and degree of reactivity within the system and with the environment.

In the petroleum upstream sector, emulsions are either used as potential solutions for various drilling and production operations or are as a nuisance in the form of oil and water mixtures during subsurface production of hydrocarbons, which requires extensive time and money for separation of oil, and water phases before transportation of hydrocarbon from the field [3].

2. Type of emulsions and their applications/occurrences in oil field

The emulsions found are of various types with respect to the droplet size, dispersion phase, and type of emulsifier used.

2.1 Emulsion as per droplet size

Emulsions are generally divided into three types with respect to the droplet size, namely, nanoemulsions (1–100 nm), microemulsions (100–400 nm), and macroemulsions (1–100 μm) [4, 5]. The droplet size influences the rheology just as it governs the stability of the emulsion by preventing coalescence and flocculation. Smaller droplet size leads to higher viscosity because of an increase in interfacial area, better elasticity, and shear thinning behavior due to the ability to deform under high stress and reform upon stress release.

2.2 Emulsion as per dispersion phase

The flow and deformation behavior (rheology) of an emulsion depends on both the nature of the droplets (dispersed phase) and the surrounding liquid (continuous phase). Moreover, the water–oil ratio will govern the flow of the emulsion. Macroemulsions (coarse emulsions) are categorized as water-in-oil (W/O) or oil-in-water (O/W) based on the continuous phase. Microemulsions, distinct from coarse emulsions, exhibit phase sensitivity and can exist in three phases: oil-in-water-in-oil (O/W/O) or water-in-oil-in-water (W/O/W) [6].

2.3 Emulsion as per type of emulsifier

The type of emulsifier severely impacts the rheological behavior of an emulsion. There are two types of emulsifiers generally used: solid particles, also known as Pickering particles and surfactant (liquid-based emulsifiers); the key difference between them has been shown in **Figure 1**. Both types have their own advantages and drawbacks as per different types of applications. In Pickering emulsions, solid particles at the interface create steric hindrance, resisting droplet movement and leading to higher viscosity compared to liquid emulsifiers. Under high shear rates, particles pack more tightly, further increasing viscosity and potentially leading to gel-like behavior, but this varies according to the overall emulsion composition and the particle characteristics, which can exhibit shear thinning upon increasing shear rates; however, it is comparatively lower than surfactant-based emulsions. These emulsions can exhibit elastic properties due to the network formed by the particles at the interface. On the contrary, surfactants reduce interfacial tension, allowing easier flow and generally resulting in lower viscosity compared to solid particle emulsifiers. Surfactants typically exhibit shear thinning behavior comparatively higher than Pickering emulsions, where viscosity decreases with increasing shear rate due to droplet alignment and deformation. Hence, these emulsions are more fluid-like and less elastic than those stabilized by solid particles. From a technical and economic perspective, it is feasible to incorporate solid particles as emulsifiers during drilling and production operations as it requires a lower amount of emulsifier concentration to maintain stability and viscosity [9, 10]; additionally, the surfactants generally used are prone to being toxic, carcinogenic, and allergic, and are not mostly biodegradable [11–13].

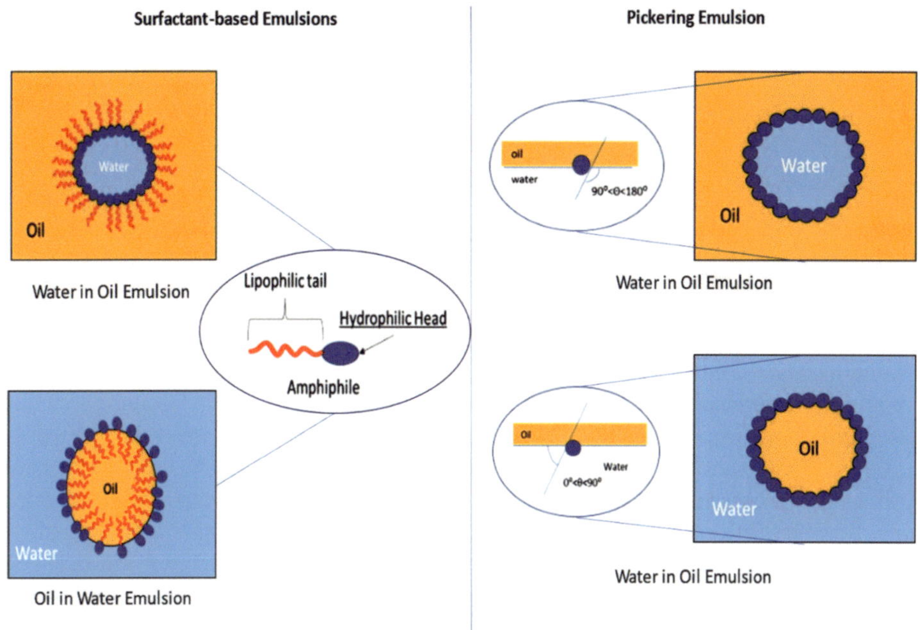

Figure 1.
The key distinction between emulsions formed with surfactants and those formed with Pickering particles. Surfactants rely on groupings of special molecules called amphiphiles and their hydrophilic–lipophilic balance (HLB) value to create different types of emulsions while for a Pickering emulsion, the particle's wettability contact angle categorizes the emulsion [7, 8].

3. Emulsion applications and occurrences in the upstream sector

Emulsions, as mentioned above, are a part of different petroleum operations, often used for a variety of applications while also acting as a hindrance during the production phase.

3.1 Drilling muds

Emulsions used in drilling applications are generally invertible emulsions that transition between O/W and W/O configurations depending on pressure and applications [14–20]. This property allows for balancing filtration control and lubricity requirements in different wellbore sections. The droplet size in these emulsions plays a significant role in their behavior and performance. Generally, a moderate droplet size range between 0.5 and 5 microns is desired. This is because traditional coarse emulsions have droplet sizes typically ranging from 1 micrometer (μm) to millimeters (mm), significantly larger than the desired range for drilling fluids. These emulsions are thermodynamically unstable and tend to separate over time, a major drawback for drilling muds requiring long-term stability; moreover, they often lack specific functionalities needed in drilling, like shale inhibition and high-temperature stability [21]. Microemulsions, although if tailor designed, can provide the required thermal stability and viscosity but are currently kinetically unstable and tend to be infeasible on a large scale [22, 23]. Smaller droplets lead to higher viscosity, improved filtration control, and potentially better shale inhibition. However, excessive viscosity can hinder flow and pumping operations. In contrast, larger droplets contribute to lower viscosity, better flow behavior, and enhanced lubricity. However, they might compromise filtration control and thus provide limited shale inhibition.

3.2 Emulsion formation during production

The emulsions produced from oilfields range from water-in-oil (W/O) or oil-in-water (O/W) to the complex Winsor type 3 (W/O/W and O/W/O)-based emulsions depending on the percentage of the phases and type of surfactants present and their affinity with the phases [24, 25] But mostly water-in-oil-based (W/O) emulsion are formed due to presence of high concentration of asphaltene in the crude oils, asphaltene is a natural surfactant which is strongly oleophilic [26, 27]. The rheology in these scenarios holds a critical role during flow assurance and refinement of the hydrocarbons, which depends on the severity of the emulsification, which is controlled by viscousness of the phases and their compositional percentage along with the temperature and pressure conditions. Emulsions significantly impact the flow behavior of produced fluids within pipelines and processing facilities. Increased viscosity in emulsions can lead to pressure drops, flow restrictions, and heat transfer issues affecting the cooling and heating of pipelines and equipment [28]. Similarly, high viscosity emulsions can complicate various downstream refining processes like separation, transportation, and processing [29].

3.3 Well stimulation (matrix acidizing)

Water-in-oil macroemulsion is synthesized for matrix acidizing operations as it can hold large quantities of acid content while utilizing low concentrations of emulsifier, and the oil phase provides retardation against reactivity between acid and the rock surface [30, 31]. **Figure 2** presents the large adaptation of water-in-oil (W/O) emulsions favored by various research studies and field cases. **Figures 3** and **4** provide a comparison between flow of a conventional acid and an emulsified acid. Conventional acid is consumed entirely halfway due to lack of a diffusion barrier to subside the acid reactivity in one area, whereas, as with emulsified acid, it successfully reaches the targeted depth due to the presence of the oil phase acting as a retarding agent between the acid and the rock surface thereby slowing down acid reactivity and minimizing consumption. In order to perform a successful acidizing operation, the shear thinning under high stress of the emulsion during propagation through the wellbore to the reservoir and the ability to thicken as the pressure subsides is crucial to be optimized while maintaining stability against thermal and mechanical stresses [32].

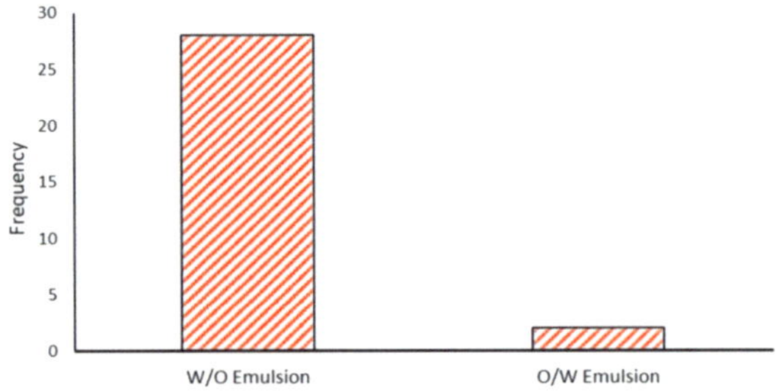

Figure 2.
The high usage of water-in-oil (W/O) emulsions which provide better retardation and sweep efficiency during acid stimulation.

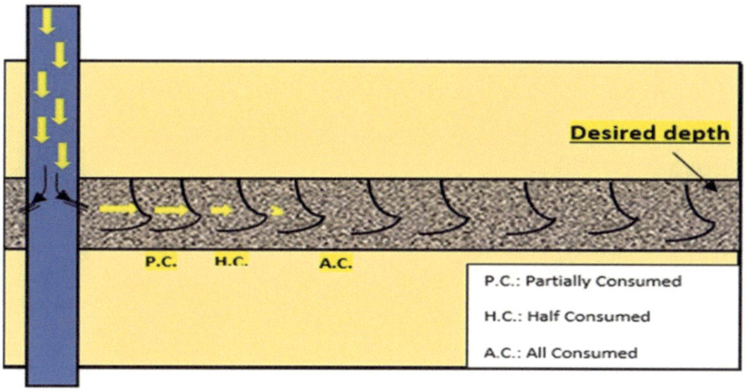

Figure 3.
How a typical acid treatment without a retarder penetrates the rock. As illustrated, the acid is completely used up before reaching even half the intended depth [7].

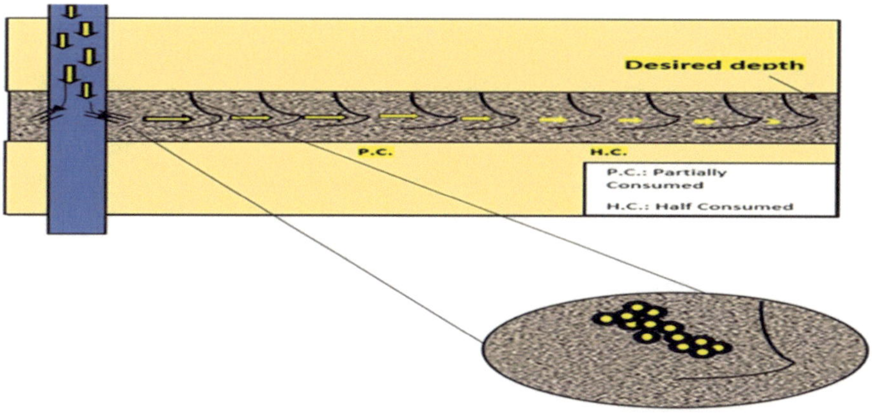

Figure 4.
The propagation of an emulsified acid that reaches the targeted depth due to oil acting as the external phase as it retards the reactivity between the acid and rock surface [7].

3.4 Conformance control

Similar to matrix acidizing, the shear thinning and thickening behavior of the emulsion is important for improving oil recovery through reservoir profile enhancement and chemical flooding at varying (low to high) temperature and pressure conditions. Contrary to the well-stimulation operations, which utilize oil as the external phase for retarding acid, the conformance control emulsions utilize oil-in-water (O/W) emulsions, as shown in **Figure 5**. This is because O/W emulsions possess adequate viscosity and hence better shear-thinning rheology, which is better suited for injectivity contrary to the comparatively more viscous W/O emulsion, which is harder to inject and induce large pressure drop during displacement [28, 33, 34]. This type of rheology generates low chances of formation damage and makes it easy for blockage removal, along with better power over droplet size, providing a wider range of droplets and making it more possible to plug different pore throats during conformance control [35–38]. **Figure 6** presents the three different type of plugging mechanisms adapted by emulsion droplets during conformance control.

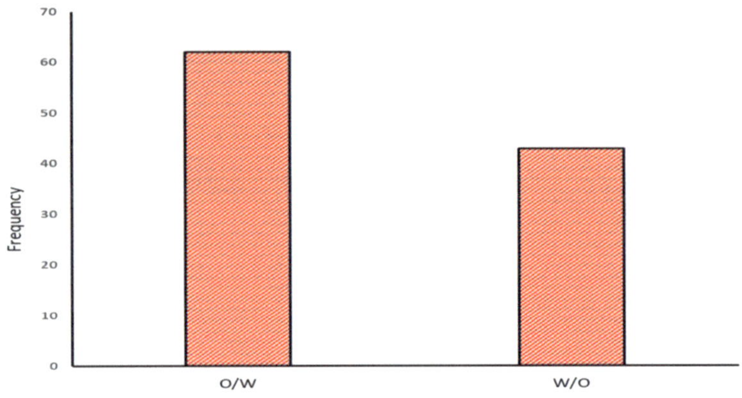

Figure 5.
The number of publications utilizing W/O and O/W emulsions for conformance control. O/W emulsions are favored more due to the provision of desired viscosity and shear thinning rheology than W/O emulsions, which have injectivity constraints and generate large pressure drops during flow.

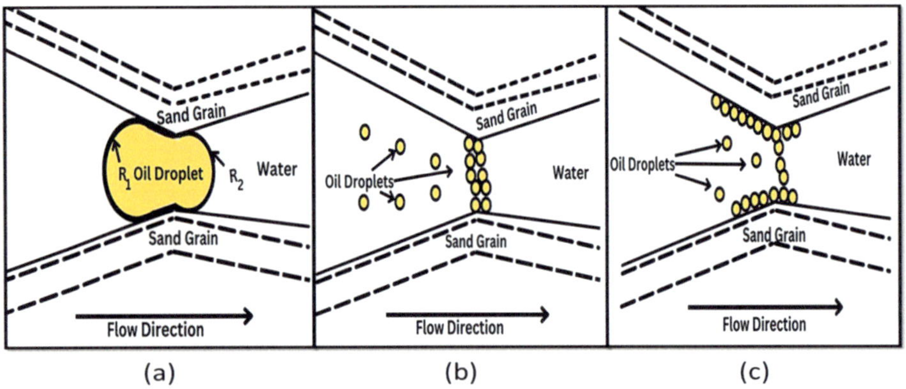

Figure 6.
The plugging mechanism that works for emulsions during conformance control: (a) In oil-in-water emulsions, as an oil droplet enters a narrow pore throat, the curved front of the droplet faces more resistance than the flatter back. This difference in resistance creates a pressure that helps the droplet lodge itself in the pore, blocking it; this is also known as the Jamin effect. (b) Additionally, emulsions can plug pores by having droplets of various sizes accumulate at the pore throat. This variety in droplet size allows them to fit into and block pores of different diameters, creating a more effective overall blockage. (c) Finally, the charged Pickering particles used in these emulsions can be attracted to the oppositely charged rock surface within the pores. This attraction helps the droplets adhere to the pore walls, further contributing to the plugging effect [8].

3.5 Enhance oil recovery

Highly permeable formations favor O/W emulsions due to their lower viscosity and better flow characteristics. Conversely, W/O emulsions might be preferred in low-permeability reservoirs to enhance sweep efficiency. Oil-wet reservoirs benefit from W/O emulsions, which can change the rock surface wettability toward water, mobilizing trapped oil. Water-wet reservoirs might find better success with O/W emulsions. Macroemulsions are sought over microemulsions generation due to high expenditure requirements attached to the smaller droplet size emulsions regardless of providing better thermal stability and, due to higher viscosity, the ability to mobilize oil from tight zones and improve sweep efficiency [39, 40].

4. Factors affecting emulsion rheology

4.1 Internal phase factors

4.1.1 Water: oil ratio

The water-to-oil ratio influences the viscosity-to-mobility ratio, which governs the overall rheological performance of an emulsion. Slight changes in a certain phase's ratio can change the continuous and discrete phases of the emulsion and may distort the affinity of the emulsifier with a certain phase. For instance, emulsions used for conformance control and acid stimulation require a balanced viscosity to mobility ratio, which provides low injection pressure requirements, controlled shear thinning and thickening behavior with great sweep efficiency for this purpose generally, 50:50, 70:30, and even 80:20 are considered as water to oil ratios respectively as this allows aqueous medium to govern the emulsion viscosity no matter how viscous the oil phase. If the oil phase has a percentage greater than 50%, then the emulsion viscosity will resemble oil viscosity, which in most cases is troublesome due to high power injection rates, along with other constraints such as higher chances of formation damage, difficulty in damage removal, stagnancy in plugging and stimulation operations as well as increasing the overall operational cost [6, 28, 35–38].

4.1.2 Droplet size distribution

Generally, smaller droplet size leads to higher viscosity in emulsions. This is because smaller droplets create a larger total interfacial area within the emulsion, requiring more energy for the fluid to flow and increasing resistance to movement [41]. However, it is crucial to account for the interplay between droplet size and the viscosity of the individual phases. If the dispersed phase is much more viscous than the continuous phase, a smaller droplet size can significantly increase viscosity. If the viscosity difference is smaller, the impact of droplet size on overall viscosity might be less pronounced. Droplet size depends on the viscosity and polarity of the individual phases, emulsifier concentration (in case of Pickering emulsions, the size of the particle is also accounted for), method of homogenization, which includes mixing speed, ultrasonication power and duration, temperature, droplet dispersion mode. Externally, the reservoir temperature, pressure, salinity, and pH of the rock mineralogy and reservoir fluid can influence the droplet size, thereby changing the flow behavior of an emulsion. For example, if the droplet size distribution is disturbed by an increase in salinity in the system, it may lead to instability mechanisms such as coalescence and flocculation. This larger droplet size can reduce the shear and thermal resistance of an emulsion due to large droplets requiring lesser energy to propagate within the continuous phase and being more sensitive to thermal conductivity [42]. This salinity increase can also have an opposite effect by providing an enhanced particle network surrounding the droplets, which depends upon the affinity between the emulsifier, the salt ions, and the phases. **Table 1** and **Figure 7** present the different emulsion droplet size ranges for emulsified acids used for matrix acidizing and emulsions used as conformance agents. The droplet size can vary depending on the specific reservoir. Factors that influence droplet size include the size of the Pickering particles used, the viscosity and polarity of the oil and water phases, the method used to create the emulsion, and the conditions within the reservoir itself [8].

Authors	Year	Droplet size	Emulsion type
Guidry et al. [43]	1989	20 μm HCl, 200 μm N_2	Macroemulsion
Al-Anazi et al. [44]	1998	77 μm HCl	Macroemulsion
Navarrete et al. [45]	1998	1–77 μm HCl	Macroemulsion
Nasr-el-din et al. [46]	2008	35 μm HCl	Macroemulsion
Zakaria and Nasr-El-Din [47]	2015	1.14–6.34 μm HCl	Macroemulsion
Sidaoui et al. [48]	2016	1.47–3.09 μm HCl	Macroemulsion
Hoefner and Fogler [49]	1987	10 nm (micelle state)	Microemulsion
Zhang et al. [50]	2008	54.5 nm	Microemulsion
Tupã et al. [51]	2016	80–183 nm	Microemulsion

Table 1.
Highlights the different droplet sizes of the emulsified acids utilized for well stimulation [7].

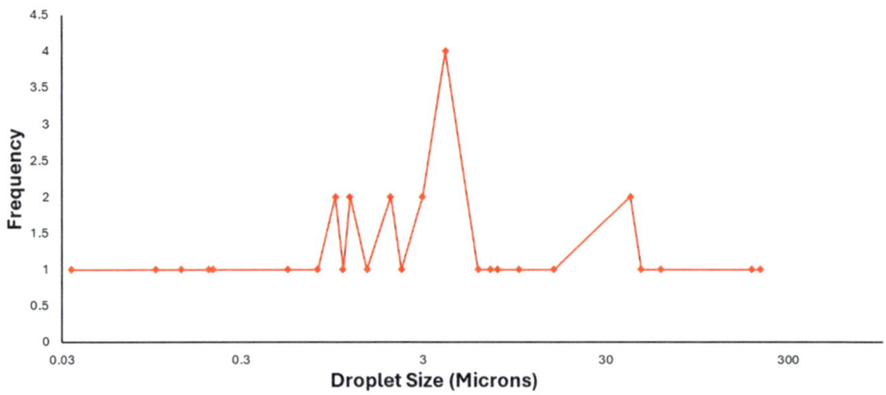

Figure 7.
A representation according to past research on the size of droplets used for conformance control which typically falls between 1 and 4 microns.

4.1.3 Oil viscosity

The flow of emulsion can largely be affected by the viscosity of the oil phase, whether it is the dispersed phase or continuous phase. Oil should be moderately viscous as per the operational conditions, so that whether oil is the dispersed or continuous medium, it can slow down the droplet propagation toward one another, thereby preventing coalescence. The emulsion will be able to exhibit the necessary flow sweep efficiency, the shear thinning behavior will be customized to provide minimum injectivity concerns, and it is useful in cases where minimum contact between droplet and rock surface is required. If the oil is highly viscous, it shows solid-like behavior leading toward emulsion destabilization, high pumping requirement, and possible formation damage [52, 53].

4.1.4 Effect of main constituents and additives on emulsion performance

The rheological performance of an emulsion, meaning its flow and deformation behavior, depends heavily on the composition of its main constituents and the presence of any additives. The viscosity, size distribution, and polarity of the dispersed phase significantly impact the overall viscosity and flow behavior of the emulsion. Smaller droplets and higher viscosity of the dispersed phase generally increase resistance to flow. Similarly, the viscosity of the continuous phase plays a major role, with higher viscosity leading to thicker emulsions. Additionally, its compatibility with the dispersed phase and emulsifier influences stability and flow characteristics. The emulsifier controls the interfacial tension between phases, promoting droplet dispersion and stability [54]. The type and concentration of emulsifier directly affect droplet size, viscosity, and flow behavior. Additives such as thickeners and thinners alter the viscosity for various purposes, controlling flowability, improving suspension of solids, or enhancing stability; easier pumping or flow is often used to compensate for the high viscosity of the emulsion. The magnitude and shear dependence of viscosity modification by additives depends on both their specific mechanism of action and their compatibility with the emulsion components. As a result, the optimal type and concentration of an additive will vary based on the desired effect and the specific properties of the emulsion [2, 55, 56].

4.1.5 Interfacial affinity /wettability of the emulsifier toward the phases

As mentioned prior, the effect on droplet size affects emulsion stability and, thereby, viscosity, the affinity between the emulsifier and the individual phases can affect how robust an emulsion will be under different temperature, shear, and salinity conditions in case of surfactant-based emulsion where the surfactant hydrophilic–lipophilic balance (HLB)-based affinity determines the governing phase it is crucial to attain a balance between both the phases (oil and water) to achieve long-term stability as required for an application [57]. Similarly, in Pickering emulsions, the solid particle wettability toward the oil and water phases is the determinant of stability, where generally partial wettability (90° contact angle) by the emulsifying particle is favored for both phases as it reduces the amount of energy required to create a steric barrier by the particle at oil–water interface [58–60]. This strengthened stability creates an emulsion that is able to flow fluently under varying shear and thermal stresses due to the strong affinity-induced network between the emulsifier and the phases. The external wettability of the emulsion with the rock surface is important to consider as flow performance gets hindered in cases of unwanted frictional resistance and slipping behavior; therefore, the viscosity and polarity of the emulsion and its individual phases need to be modified as per the reservoir mineralogy and application requirement. For instance, during conformance control, plugging performance is highly affected not just by the wettability between the emulsion and the rock during flow but also depends on the elasticity of the inner droplet network (in case of Pickering emulsion interparticle network as well), which is highly dependent on the emulsifier interaction with the phases [8, 52].

4.1.6 Density

The density of individual phases is vital for a stable emulsion, which affects the emulsion viscosity. A large density difference between two phases can lead to fast gravity segregation-based destabilization (creaming and sedimentation), whereas a low-density difference can prevent this phenomenon from occurring. Highly dense emulsions tend to be generally highly viscous due to the increased mass per unit volume, although other factors such as droplet size, temperature, gravitational forces, viscosity of individual phases, emulsifier type, and shear rate. The impact of density is often more pronounced in dilute emulsions with a significant phase difference. In densely packed emulsions, the overall density might dominate the individual phase densities. The density difference can affect the homogenization process used to create the emulsion. This, in turn, influences the droplet size distribution, which significantly impacts viscosity and flow behavior [8].

4.2 External phase factors

4.2.1 Reservoir (mineralogy, and fluid pH and salinity) composition and heterogeneity

Reservoir mineralogy and its fluid composition hold content with different salinity and polarities, which when interacting with the injected emulsion, can either increase the stability and viscosity of the emulsion or cause emulsion disassociation. These invasive ions can either create an additional particle network around the droplets along with the emulsifiers to strengthen the interfacial film between the dispersed and continuous phase, or these ions can draw the weakly bonded emulsifier ions toward themselves, causing destabilization of the interfacial film. The increase in stability does generate an increase in viscosity due to the additional particle volume and structure of the interfacial film; the additional particles in liquid increase their resistance to flow. The magnitude of viscosity increase depends on the thickness and rigidity of the film, the volume fraction of the dispersed phase, and the initial viscosity of the phases. Therefore, the emulsion is optimized during the design phase according to the mineralogy and fluid constituents of the reservoir so that the stability remains intact, and the emulsion causes minimal concerns during the displacement profile [61, 62].

4.2.2 Temperature

Temperature plays a key role in emulsion stability, much like pH and salt levels. It affects the viscosity of the liquids, which in turn influences the size and surface tension of the droplets. Similar to how pH and salt alter the ionic arrangement of the emulsion (colloid), during synthesis, high temperatures can break down droplets into smaller ones and energize the ionic charges because of an increase in thermal energy [63]. However, the specific impact of temperature depends on the chosen emulsifier and the nature of the liquids involved. During propagation in the subsurface, high temperatures and salinity can lead to either stabilization or destabilization of the emulsion. Emulsions, whether naturally occurring or intentionally introduced, need to be resistant to the high temperatures encountered in reservoirs, typically ranging from 80 to150°C [64–70].

4.3 Additional factor

4.3.1 Aging and stability overtime

As the emulsion ages, it heads toward destabilization due to several mechanisms that affect the rheology by disturbing the internal structural integrity of the colloidal system. Mechanisms influenced by gravity segregation and density differences such as creaming and sedimentation affect the composition and flow behavior by separation of the phases heading chromatographic fractionation, having each individual phase moving at a different speed, having a different inclination toward reservoir fluids and mineralogy. This, in certain cases, can be harmful if unintended such as during matrix acidizing [46], but in conformance control, if designed carefully, can provide a beneficial plugging mechanism such as during simultaneous oil production and water plugging through a production zone [71].

5. Rheology testing methods

The significance of rheometry in analyzing emulsions cannot be overstated, as it sheds light on both their stability and their behavior under flow. Viscosity serves as a key indicator of an emulsion's resistance to instability, with higher viscosities reflecting a lower tendency for droplets to intermingle and destabilize due to retarded fluidity. Rheometry delves deeper by quantifying the mobility–viscosity ratio, offering valuable insights into both the efficiency of mixing during synthesis and the injection requirements for deployment. This ratio captures the delicate balance between viscosity, which resists flow, and mobility, which facilitates movement. Grasping this interplay is crucial for optimizing processes like emulsion mixing and injection into reservoirs. Rheology of drilling fluids affects hole cleaning, pressure control, and wellbore stability. Rheometry is crucial in optimizing the emulsion stability and viscosity for controlled acid release, reduced corrosion, and better penetration into the formation during matrix acidizing applications. During enhanced oil recovery operations, the viscosity and flow behavior of injected emulsions significantly influences sweep efficiency and oil displacement. While in flooding control, where redistributing flows within a reservoir take place, emulsion slugs are used as diverting agents to temporarily plug high permeability zones and divert flow toward less permeable zones.

Beyond pure viscosity, it is important to acknowledge that shear thinning behavior, where the viscosity decreases with increasing shear rate, is common in emulsions. This can be advantageous for pumping and injection but necessitates consideration of varied shear conditions. Additionally, some emulsions exhibit a yield stress, requiring a minimum shear stress to initiate flow. This factor becomes relevant in applications with low shear forces. Both viscosity and flow behavior can be temperature-dependent, making rheometry under application-relevant temperatures essential for accurate predictions.

5.1 Rotational rheometry

Emulsions possess shear thinning behavior and rotational viscometry can cover a wide range of shear rates to capture this behavior accurately. A sample is subjected to a controlled shear rate and shear stress through which the angular viscosity of the sample can then be measured. This is useful for emulsions used in drilling and production. During drilling, rotational rheometry is used to characterize viscosity, shear thinning, and yield stress of drilling fluid emulsions under simulated downhole conditions, including temperature and pressure. In the application of matrix acidizing, where emulsified acids are utilized, rotational viscometry helps ensure proper emulsification and stability of the acid during pumping and placement in the reservoir. For enhanced oil recovery where mobility improvement is undertaken, rotational viscometry helps design emulsions with tailored viscosity and shear thinning properties for better mobility in porous media and improved oil recovery. For conformance control operations enhancement of reservoir profile, rotational rheometric technique is used to optimize emulsions with desired viscosity, gelation properties, and shear thinning behavior for effective plugging and subsequent flow through less permeable zones.

5.2 Oscillatory rheometry

In dynamic rheometry, a sample is subjected to an oscillating shear stress, and the strain response is measured. The storage modulus (G′) and loss modulus (G″) can then be calculated from the strain response. The storage modulus is a measure of the elastic behavior of the sample, while the loss modulus is a measure of the viscous behavior of the sample. Here are two main types of tests carried out in oscillatory rheometry, namely, (a) amplitude sweep and (b) frequency sweep [72, 73].

a. Strain (Amplitude) Sweep: Here, the frequency (ω) is held constant, and the amplitude of the strain (γ_0) is gradually increased. This helps identify the linear viscoelastic region (LVR), where the stress response remains proportional to the strain and independent of the amplitude.

b. Frequency Sweep: Once the LVR is identified, the amplitude of the strain (γ_0) is kept constant, and the frequency (ω) is varied. This test measures the viscoelastic properties of the sample, such as the storage modulus (G′) and the loss modulus (G″), throughout a range of frequencies.

Generally, rotational rheometry requires different geometries like cone-and-plate, parallel plates, and concentric cylinders, depending on sample volume and viscosity. While Oscillatory rheometers might have additional features like temperature control and wider frequency ranges compared to rotational ones. Often, specialized types (e.g., microfluidic rheometers) might be specific to either rotational or oscillatory measurements [74].

For reservoir condition-based applications, it is safe to state that for both the rheometric techniques, hybrid rheometers can be employed, which utilize diverse plate geometries to analyze the rheology of various fluids under varying temperature conditions and shear stresses [75, 76]. This adaptability is crucial to account for the unique behaviors of different fluids, such as slipping and phase evaporation under high heat. For highly viscous samples, parallel plates offer suitable measurement, while less viscous colloids benefit from the enclosed environment of concentric cylinders. The expansive surface area of parallel plates can lead to potential fluid slippage and evaporation in low-viscosity media, an issue less significant for their highly viscous counterparts. Conversely, concentric cylinders effectively confine the fluid, minimizing such concerns even with elevated temperatures. This versatility in plate geometry allows rheometers to accurately assess the rheological properties of a wide range of fluids, ensuring reliable data across diverse applications. **Figure 8** shows (a) Searle-type viscometer and (b) Couette-type viscometer. In a Searle-type viscometer, the viscosity measurement can be performed by rotating the inner cylinder as the resulting torque is measured on the rotating tool, which is usually the bob, the cone, or the upper plate. In contrast, a Couette-type viscometer is used to rotate the outer cylinder to take viscosity measurements because the resulting torque is measured on the counterpart of the rotating tool [77, 78].

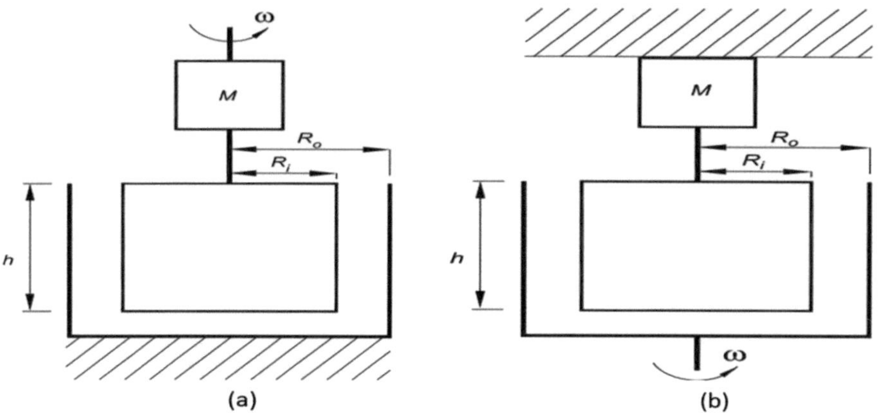

Figure 8.
(a) Searle-type viscometer (viscosity measurement can be performed by rotating the inner cylinder) and (b) Couette-type viscometer (viscosity measurement can be performed by rotating the outer cylinder). Where ω = angular velocity in radians per second, M = torque in newton-meters acting on the cylinder surface, Ri = radius in meters of the inner cylinder, Ro = radius in meters of the outer cylinder, h = height of immersion in meters of the inner cylinder in the liquid medium.

5.3 Extensional Rheometry

In extensional rheometry, a sample is subjected to an extensional flow, and the extensional viscosity is measured. Extensional viscosity is a measure of the resistance of a material to stretching. Extensional flow analysis can help optimize emulsion stability to prevent breakdown and ensure proper mud rheology during drilling operations. This is conducted to better understand how emulsions interact with cuttings and solids during hole cleaning under various flow conditions [79]. Design emulsions with tailored extensional properties (enables the development of emulsions with improved resistance to extensional forces for better performance) for specific drilling environments and challenges [80]. Although it is more targeted toward polymer and surfactant flooding during enhanced oil recovery [81, 82], in the case of emulsion utilization in the application, this technique can help in perceiving how emulsions interact with rock minerals and pore structures, influencing oil mobilization mechanisms. Provide multiple insights regarding changes in viscosity and droplet deformation, the effect of extensional forces encountered during flow through a porous medium, which can help in designing emulsions with optimal viscosity and extensional behavior for efficient flow and sweep for specific reservoir characteristics and target zones to maximize oil recovery. Similarly, for conformance control, the effect of extensional forces during emulsion propagation through different permeable zones can help analyze droplet deformation and flow behavior, influencing diversion effectiveness and plugging performance and allowing for developing diverting agents with better adaptability to extensional forces.

5.4 Microfluidic rheometry

Microfluidic chips have recently been adapted vastly to study steady shear viscosity for enhanced oil recovery and mobility control applications [6, 83–88].

These miniature reservoir models allow for a better understanding of the fluid flow and optimization on a microscale, mimicking specific reservoir geometries for targeted analysis. These models allow for testing numerous emulsion formulations under different temperature, pressure, and salinity conditions with limited expense of resources. For enhanced oil recovery and conformance control applications, this technique has permitted the development of emulsions with optimal viscosity and mobility for efficient sweep efficiency. Study of emulsion-rock interactions to understand mechanisms of oil mobilization and entrapment. Design diverting agents with better performance for effective plugging of high permeability zones while observing the colloidal systems flow behavior in the presence of diverse reservoir heterogeneity.

Microfluidic scale rheometry is an emerging technique [89] which is currently utilized mostly for testing extensional rheology of polymer-based fluids in enhanced oil recovery applications [90–93]. However, it has the potential to optimize drilling muds and analyze small sample volumes, which is crucial for limited availability or expensive drilling fluids. Generate data at high shear rates encountered during drilling, enabling better prediction of flow behavior, which can be used to improve viscosity and shear thinning behavior for efficient hole cleaning and pressure control. Examine fluid-particle interactions to understand suspension stability and prevent clogging. Emulsified acids can be designed more efficiently by implementing microfluidic techniques. These will help analyze small volumes of the emulsion relevant to lab-scale development and optimization. Simulate flow through porous media within the microfluidic chip, mimicking acid transport in the reservoir. This will enable better investigation of reaction kinetics and interactions between acid and rock formations at a finer scale, thereby improving the design of emulsions with controlled acid release and improved penetration depth. Moreover, this can work to strengthen emulsion stability and minimize corrosion.

Microfluidic holds multiple advantages, from enabling rapid screening and optimization of numerous emulsion formulations to reducing material requirements and facilitating research with finite resources. Simulate controlled environments to replicate specific downhole conditions, which can be combined with visualization tools for a deeper understanding of flow behavior. There are also limitations associated with scaling challenges as well; transforming results from microfluidic chips to larger-scale field applications requires multiple factors to be amended. These small-scale models sometimes allow for limited geometry options to mimic the complexity of real-world porous media. The equipment might not be readily available in all settings. Despite these current limitations, it is an emerging field that presents a valuable tool for future research and development. Investigating emulsions at relevant scales and simulating flow conditions can contribute to designing and deploying more effective emulsions for optimal well management, stimulation, and enhanced oil recovery strategies. **Figure 9 a** and **b** presents a working principal diagram of a microfluidic chip rheometer and a graphical representation of changes in flow rate, shear rate, and viscosity upon the position of the two fluids on the microfluidic chip. This technique measures a sample's viscosity by injecting it alongside a reference fluid into a microfluidic chip. By precisely controlling the flow rates and capturing images of the merging streams, the position where the two fluids meet (interface) is precisely measured. Since a change in the sample's viscosity alters the space it occupies within the chip, the interface position will also shift. This relationship between flow rates, occupied volume, and the interface location allows for a straightforward calculation of the sample's viscosity [94].

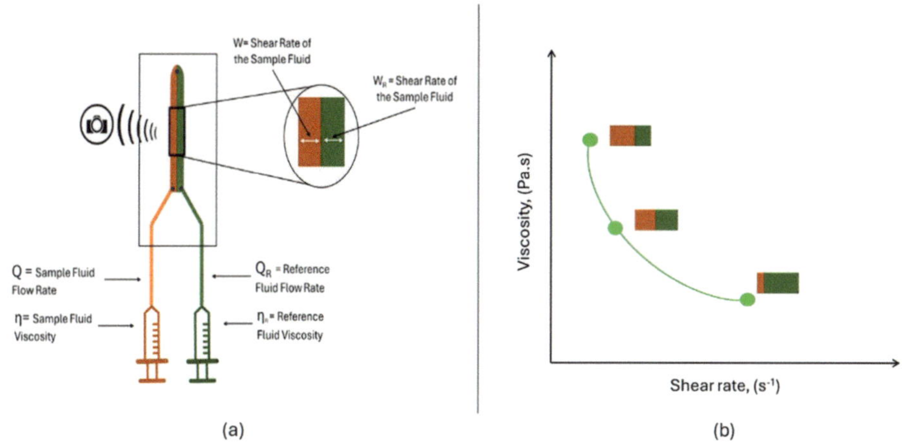

Figure 9.
(a) shows a working illustration of a microfluidic chip rheometer, and (b) presents a graphical representation of how varying the flowrate changes the shear rate and the viscosity of the fluids, thereby changing the fluid position on the interface, which is captured by the microscopic camera set on top of the chip.

6. Conclusion

Understanding emulsion rheology is crucial for optimizing their performance in various upstream petroleum operations. This chapter has explored the role of emulsion rheology in various upstream petroleum operations. By tailoring the composition and properties of emulsions, operators can achieve improved wellbore stability, enhanced oil recovery, and efficient fluid flow management. The emerging trends such as adaptation of solid particles as emulsifiers, need to be explored further to make them more resistant to varying thermal and shear disturbances during the fluid flow from the surface to the reservoir. Similarly, these enhancements can be tested and optimized more economically and efficiently using microfluidic scale rheometers, allowing for more effective rheological optimization at a finer scale. Considering the implementation of these trends can lead to more productive outcomes in the petroleum fields, which, as per the current scenario, require inexpensive solutions, especially during the designing and deployment stages [1].

Acknowledgements

The authors would like to acknowledge the funding and support from the following Yayasan Universiti Teknologi PETRONAS grants: YUTP 015LC0-335 "Mechanistic Investigation of Pickering Emulsified Polymeric Gel for Water Shut-off Conformance Control" and YUTP 015LC0-406 "New Sand Screen Design for Mitigating Common Failure Mechanisms in Water Injection Wells".

Author details

Iskandar Bin Dzulkarnain[1*], Muhammad Mohsin Yousufi[2]
and Mysara Eissa Mohyaldinn Elhaj[3]

1 Centre of Reservoir Dynamics, Institute of Sustainable Energy, Universiti Teknologi PETRONAS, Malaysia

2 Petroleum Engineering Department, Universiti Teknologi PETRONAS, Malaysia

3 Centre of Flow Assurance, Institute of Sustainable Energy, Universiti Teknologi PETRONAS, Malaysia

*Address all correspondence to: iskandar_dzulkarnain@utp.edu.my

References

[1] Khan BA, Akhtar N, Khan HMS, Waseem K, Mahmood T, Rasul A, et al. Basics of pharmaceutical emulsions: A review. African Journal of Pharmacy and Pharmacology. **2011**;5(25):2715-2725

[2] Tadros TF. Emulsions: Formation, Stability, Industrial Applications. Berlin, Germany: De Gruyter; 2016. DOI: 10.1515/9783110452242-002

[3] Hasan SW, Ghannam MT, Esmail N. Heavy crude oil viscosity reduction and rheology for pipeline transportation. Fuel. 2010;**89**(5):1095-1100

[4] Souto EB, Cano A, Martins-Gomes C, Coutinho TE, Zielińska A, Silva AM. Microemulsions and nanoemulsions in skin drug delivery. Bioengineering. 2022;**9**(4):1-22. DOI: 10.3390/bioengineering9040158

[5] Sharma MK, Shah DO. Introduction to Macro- and Microemulsions. Washington, D.C.: American Chemical Society; 1985. pp. 1-18. DOI: 10.1021/bk-1985-0272.ch001

[6] Rezaei N, Firoozabadi A. Macro- and microscale waterflooding performances of crudes which form W/O emulsions upon mixing with brines. Energy and Fuels. 2014;**28**(3):2092-2103. DOI: 10.1021/ef402223d

[7] Yousufi MM, Eissa M, Elhaj M, Dzulkarnain IB. A review on use of emulsified acids for matrix acidizing in carbonate reservoirs. ACS Omega. 2024;**9**(10):11027-11049. DOI: 10.1021/acsomega.3c07132

[8] Yousufi MM, Dzulkarnain I, Eissa M, Elhaj M, Ahmed S. A perspective on the prospect of pickering emulsion in reservoir conformance control with insight into the influential parameters and characterization techniques. Processes (MDPI). 2023;**11**:262, 1-29. DOI: 10.3390/pr11092672

[9] Al-Yaari M, Hussein IA, Al-Sarkhi A. Pressure drop reduction of stable water-in-oil emulsions using organoclays. Applied Clay Science. 2014;**95**:303-309

[10] Stavland A, Andersen KI, Sandoey B, Tjomsland T, Mebratu AA. How to apply a blocking gel system for bullhead selective water shutoff: From laboratory to field. In: SPE/DOE Symposium on Improved Oil Recovery. OnePetro; 2006

[11] Hossain KMZ, Deeming L, Edler KJ. Recent progress in pickering emulsions stabilised by bioderived particles. RSC Advances. 2021:39027-39044. DOI: 10.1039/d1ra08086e

[12] Yang Y, Fang Z, Chen X, Zhang W, Xie Y, Chen Y, et al. An overview of pickering emulsions: Solid-particle materials, classification, morphology, and applications. Frontiers in Pharmacology. 2017;**8**:1-20. DOI: 10.3389/fphar.2017.00287

[13] Presley CL, Militello M, Barber C, Ladd R, Laughter M, Ferguson H, et al. The history of surfactants and review of their allergic and irritant properties. Dermatitis. 2021;**32**(5):289-297. DOI: 10.1097/DER.0000000000000730

[14] Marinescu P, Young S, Iskander GR. New generation ultra-high temperature synthetic-based drilling fluid; development and best practices for extreme conditions and ECD management. In: Abu Dhabi International Petroleum Exhibition and Conference. Richardson, TX, USA: SPE; 2014. p. D021S033R002

[15] Srungavarapu M, Patidar KK, Pathak AK, Mandal A. Performance studies of water-based drilling fluid for drilling through hydrate bearing sediments. Applied Clay Science. 2018;**152**:211-220

[16] He Y, Long Z, Lu J, Shi L, Yan W, Liang D. Investigation on methane hydrate formation in water-based drilling fluid. Energy & Fuels (Washington D.C.). 2021;**35**(6):5264-5270

[17] Nikolaev NI, Tianle L, Zhen W, Guosheng J, Jiaxin S, Mingming Z, et al. The experimental study on a new type low temperature water-based composite alcohol drilling fluid. Procedia Engineering. 2014;*73*:276-282

[18] Taugbøl K, Fimreite G, Prebensen OI, Svanes K, Omland TH, Svela PE, et al. Development and field testing of a unique high-temperature/high-pressure (HTHP) oil-based drilling fluid with minimum rheology and maximum sag stability. In: SPE Offshore Europe Conference and Exhibition. Aberdeen, United Kingdom: SPE; 2005. p. SPE-96285

[19] Zhao S, Yan J, Shu Y, Zhang H. Rheological properties of oil-based drilling fluids at high temperature and high pressure. Journal of Central South University of Technology. 2008;**15** (Suppl. 1):457-461

[20] Xu Z, Song X, Li G, Zhu Z, Zhu B. Gas kick simulation in oil-based drilling fluids with the gas solubility effect during high-temperature and high-pressure well drilling. Applied Thermal Engineering (Amsterdam, Netherlands: Elsevier). 2019;**149**:1080-1097

[21] Fakoya MF, Ahmed RM. A generalized model for apparent viscosity of oil-based muds. Journal of Petroleum Science and Engineering. 2018;**165**(March):777-785. DOI: 10.1016/j.petrol.2018.03.029

[22] Fluid MD, Pine OW, Sousa RPFD, Braga GS, Silva RR, Leal GLR, et al. Formulation and study of an environmentally friendly microemulsion-based drilling fluid (O/W) with pine oil. Energies. 2021;**14**(23):1-22

[23] Silva RR, Garnica AIC, Leal GLR, Viana LR, Júlio C, Neto DS, et al. Evaluation of novel microemulsion-based (O/W) drilling fluid with nonionic surfactant and shale interaction mechanisms. Journal of Petroleum Science and Engineering. 2022;**213**(March):1-16. DOI: 10.1016/j.petrol.2022.110327

[24] Ruckenstein E. Microemulsions, macroemulsions, and the Bancroft rule. Langmuir. 1996;**12**(26):6351-6353. DOI: 10.1021/la960849m

[25] Gómora-figueroa AP, Camacho-velázquez RG, Guadarrama-cetina J, Guerrero-sarabia TI. Oil emulsions in naturally fractured porous media. Petroleum. 2019;**5**(3):215-226. DOI: 10.1016/j.petlm.2018.12.004

[26] McLean JD, Kilpatrick PK. Effects of ASPHALTENE solvency on stability of water-in-crude-oil emulsions. Journal of Colloid and Interface Science. 1997;**189**(2):242-253. DOI: 10.1006/jcis.1997.4807

[27] Kilpatrick PK. Water-in-crude oil emulsion stabilization: Review and unanswered questions. Energy and Fuels. 2012;**26**(7):4017-4026. DOI: 10.1021/ef3003262

[28] Lim JS, Wong SF, Law MC, Samyudia Y, Dol SS. A review on the effects of emulsions on flow behaviours and common factors affecting the stability of emulsions. Journal of Applied Sciences. 2015;**15**(2):167-172

[29] Wong SF, Lim JS, Dol SS. Crude oil emulsion: A review on formation, classification and stability of water-in-oil emulsions. Journal of Petroleum Science and Engineering. 2015;**135**:498-504. DOI: 10.1016/j.petrol.2015.10.006

[30] Yousufi MM, Elhaj MEM, Moniruzzaman M, Ayoub MA, Nazri ABM, Husin HB, et al. Synthesis and evaluation of Jatropha oil-based emulsified acids for matrix acidizing of carbonate rocks. Journal of Petroleum Exploration and Production Technologies. 2019;**9**(2):1119-1133. DOI: 10.1007/s13202-018-0530-8

[31] Adewunmi AA, Solling T, Sultan AS, Saikia T. Emulsified acid systems for oil well stimulation: A review. Journal of Petroleum Science and Engineering. 2022;**208**:109569. DOI: 10.1016/j.petrol.2021.109569

[32] Ahmed M, Sultan A, Qiu X, Sidaoui Z, Ali AAA. A novel emulsified acid for deep wells stimulation: Rheology, stability, and Coreflood study. In: SPE Kingdom of Saudi Arabia Annual Technical Symposium and Exhibition; Dammam, Saudi Arabia. Vol. 2018. Richardson, TX, USA: Society of Petroleum Engineers (SPE); 2018. p. 2018. DOI: 10.2118/192312-ms

[33] Zhang T, Davidson A, Bryant SL, Huh C. Nanoparticle-stabilized emulsions for applications in enhanced oil recovery. Proceedings - SPE Symposium Improved Oil Recover. 2010;**2**(April):1009-1026. DOI: 10.2118/129885-ms

[34] Li S, Yu L, Lau HC, Stubbs LP. Experimental study of mobility control by clay-stabilized pickering emulsions in high-salinity reservoirs. In: Proceedings - SPE Annual Technical Conference and Exhibition. 2020. pp. 1-15. DOI: 10.2118/201256-MS

[35] Yu L, Dong M, Ding B, Yuan Y. Experimental study on the effect of interfacial tension on the conformance control of oil-in-water emulsions in heterogeneous oil sands reservoirs. Chemical Engineering Science. 2018;**189**:165-178

[36] Jurinak JJ, Summers LE. Oilfield applications of colloidal silica gel. SPE Production Engineering. 1991;**6**(04):406-412

[37] Zhao XJ, Wang CJ. A study on microscopic profiling/oil displacing mechanisms in weak gel flooding. Oilfield Chemistry. 2004;**21**(1):56-60

[38] McAuliffe CD. Oil-in-water emulsions and their flow properties in porous media. Journal of Petroleum Technology. 1973:727-733

[39] Nazar MF, Shah SS, Khosa MA. Microemulsions in enhanced oil recovery: A review. Petroleum Science and Technology. 2011;**29**(13):1353-1365. DOI: 10.1080/10916460903502514

[40] Ahmed S, Elraies KA. Microemulsion in enhanced oil recovery. In: Karakuş S, editor. Science and Technology Behind Nanoemulsions. Rijeka: IntechOpen; 2018. DOI: 10.5772/intechopen.75778

[41] Shafiei M, Kazemzadeh Y, Martyushev DA, Dai Z, Riazi M. Effect of chemicals on the phase and viscosity behavior of water in oil emulsions. Scientific Reports. 2023;**13**(1):4100. DOI: 10.1038/s41598-023-31379-0

[42] L'Estimé M, Schindler M, Shahidzadeh N, Bonn D. Droplet size distribution in emulsions. Langmuir. 2023;**40**(1):275-281. DOI: 10.1021/acs.langmuir.3c02463

[43] Guidry GS, Ruiz GA, Saxon A. SXE/N2 matrix acidizing. In: SPE Middle East

Oil Technical Conference and Exhibition, SPE-17951-MS. Manama, Bahrain: Society of Petroleum Engineers; 1989. pp. 1-10. DOI: 10.2118/17951-MS

[44] Al-Anazi HA, Nasr-El-Din HA, Mohamed SK. Stimulation of tight carbonate reservoirs using acid-in-diesel emulsions: Field application. In: SPE International Symposium on Formation Damage Control, SPE-39418-MS. Louisiana: Society of Petroleum Engineers; 1998. pp. 9-17. Available from: https://doi.org/SPE39418-MS

[45] Navarrete RC, Holms BA, McConnell SB, Linton DE. Emulsified acid enhances well production in high-temperature carbonate formations. In: SPE European Petroleum Conference, SPE-0699-0047-JPT. Hague, Netherland: Society of Petroleum Engineers; 1998. pp. 391-406. DOI: 10.2118/0699-0047-JPT

[46] Nasr-el-din HA, Texas A, Al-dirweesh S, Aramco S. Development and field application of a new, highly stable emulsified acid. In: SPE Annual Technical Conference and Exhibition, SPE -115926-MS. Vol. 2. Denver, Colarado, USA: Society of Petroleum Engineers; 2008. pp. 1-11. DOI: 10.2118/115926-MS

[47] Zakaria AS, Nasr-El-Din HA. Application of novel polymer assisted emulsified acid system improves the efficiency of carbonate acidizing. In: SPE International Symposium on Oil Field Chemistry, SPE-173711-MS. Woodlands, USA: Society of Petroleum Engineers; 2015. pp. 1-23. DOI: 10.2118/173711-MS

[48] Sidaoui Z, Sultan AS, Qiu X. Viscoelastic properties of novel emulsified acid using waste oil : Effect of emulsifier concentration, mixing speed and temperature. In: SPE Kingdom of Saudi Arabia Annual Technical Symposium and Exhibition, SPE-182845-MS. Dammam, Saudi Arabia: Society of Petroleum Engineers; 2016. pp. 1-11. DOI: 10.2118/182845-MS

[49] Hoefner ML, Fogler HS. Role of acid diffusion in matrix acidizing of carbonates. Journal of Petroleum Technology. 1987;**39**(2):203-208. DOI: 10.2118/13564-pa

[50] Zhang S, Wang F, Chen Y, Fang B, Lu Y. Preparation and properties of diesel oil microemulsified acid. Chinese Journal of Chemical Engineering. 2008;**16**(2):287-291. DOI: 10.1016/S1004-9541(08)60076-2

[51] Tupã P, Aum P, Souza TN, Pereira YK. New acid O / W microemulsion Systems for Application in carbonate acidizing. International Journal of Advanced Science and Technology Research. 2016;**1**(6):182-196

[52] Chen Z, Dong M, Husein M, Bryant S. Effects of oil viscosity on the plugging performance of oil-in-water emulsion in porous media. Industrial and Engineering Chemistry Research. 2018;**57**(21):7301-7309. DOI: 10.1021/acs.iecr.8b00889

[53] Chesters AK. Modelling of coalescence processes in fluid-liquid dispersions. A review of current understanding. Chemical Engineering Research and Design. 1991;**69**(4):259-227

[54] Mohammed MAR, Ali S, Mohammed A. Effect of additives on rheological properties of invert emulsions. Iranian Journal of Chemistry and Petroleum Engineering. 2009;**10**(3):31-39

[55] Pal R. Rheology of simple and multiple emulsions. Current Opinion in Colloid & Interface Science (Amsterdam, Netherlands: Elsevier). 2011;**16**(1):41-60. DOI: 10.1016/j.cocis.2010.10.001

[56] Tadros TF. Fundamental principles of emulsion rheology and their applications. Colloids Surfaces A Physicochemical and Engineering Aspects. 1994;**91**:39-55. DOI: 10.1016/0927-7757(93)02709-N

[57] Griffin WC, GRIFFIN; C., W. Classification of surface-active agents by" HLB". Journal of the Society of Cosmetic Chemists. 1949;**1**(5):311-326

[58] Teo SH, Chee CY, Fahmi MZ, Wibawa Sakti SC, Lee HV. Review of functional aspects of Nanocellulose-based pickering emulsifier for non-toxic application and its colloid stabilization mechanism. Molecules. 2022;**27**:1-43. DOI: 10.3390/molecules27217170

[59] Dickinson E. Biopolymer-based particles as stabilizing agents for emulsions and foams. Food Hydrocolloids. 2017;**68**:219-231. DOI: 10.1016/j.foodhyd.2016.06.024

[60] Joseph C, Savoire R, Harscoat-Schiavo C, Pintori D, Monteil J, Leal-Calderon F, et al. O/W Pickering emulsions stabilized by cocoa powder: Role of the emulsification process and of composition parameters. Food Research International. 2019;**116**(September 2018):755-766. DOI: 10.1016/j.foodres.2018.09.009

[61] Fu L, Ma Q, Liao K, An J, Bai J, He Y. Application of Pickering emulsion in oil drilling and production. Nanotechnology Reviews. 2022;**11**(1):26-39. DOI: 10.1515/ntrev-2022-0003

[62] Li W, Jiao B, Li S, Faisal S, Shi A, Fu W, et al. Recent advances on Pickering emulsions stabilized by diverse edible particles: Stability mechanism and applications. Frontiers in Nutrition. 2022;**9**(May):1-17. Available from: https://www.frontiersin.org/articles/10.3389/fnut.2022.864943

[63] Sidaoui Z, Sultan AS. Formulating a stable emulsified acid at high temperatures: Stability and rheology study. International Petroleum Technology Conference Proceeding 2016, IPTC-19012-MS. Bangkok, Thailand: Society of Petroleum Engineers; 2016. pp. 1-17. DOI: 10.2523/IPTC-19012-MS

[64] Sidaoui Z, Sultan AS, Brady D. A novel approach to formulation of emulsified acid using waste oil. Soc. Pet. Eng. - SPE Kingdom Saudi Arabia Annual Technical Symposium and Exhibition. 2017;**2017**:1657-1680. DOI: 10.2118/188116-ms

[65] Zhou X, AlOtaibi FM, Kamal MS, Kokal SL. An experimental study on oil recovery performance using in situ supercritical CO2- emulsion for carbonate reservoirs. Society of Petroleum Engineers - Abu Dhabi International Petroleum Exhibition and Conference. 2020;**2020**:2020. DOI: 10.2118/203011-ms

[66] Sayed MA, Nasr-El-Din HA. Acid treatments in high temperature dolomitic carbonate reservoirs using emulsified acids: A Coreflood study. In: SPE Production and Operations Symposium, Proceeding. 2013. pp. 203-215. DOI: 10.2118/164487-ms

[67] Yue L, Pu W, Zhao T, Zhuang J, Zhao S. A high performance magnetically responsive Janus Nano-emulsifier: Preparation, emulsification characteristics, interfacial rheology, and application in emulsion flooding. Journal of Petroleum Science and Engineering. 2022;**208**:1-11 (PB), 109478. DOI: 10.1016/j.petrol.2021.109478

[68] Juntarasakul O, Maneeintr K. Evaluation of stability and viscosity measurement of emulsion from oil from production in northern oilfield in Thailand. In: IOP Conference

Series: Earth and Environmental Sciences. Vol. 140, No. 1. 2018. pp. 1-6. DOI: 10.1088/1755-1315/140/1/012024

[69] Wang Z, Babadagli T, Maeda N. Preliminary screening and formulation of new generation nanoparticles for stable Pickering emulsion in cold and hot heavy-oil recovery. SPE Reservoir Evaluation and Engineering. 2021;**24**(1):66-79. DOI: 10.2118/200190-PA

[70] Eghbal MD, Behzad S. Hydrophobic silica nanoparticle-stabilized invert emulsion as drilling fluid for deep drilling. Petroleum Science. 2017;**14**(1):105-115. DOI: 10.1007/s12182-016-0135-0

[71] Saikia T, Sultan AS, Hussaini SR, Barri A, Khamidy NI, Shamsan AA, et al. Application of a Pickering emulsified polymeric gel system as a water blocking agent. ACS Omega. 2021;**6**(46):30919-30931. DOI: 10.1021/acsomega.1c02956

[72] Hagar HS, Jufar SR, Lee JH, Al-mahbashi N, Alameen MB, Kwon S, et al. Chitin nanocrystals: A promising alternative to synthetic surfactants for stabilizing oil-in-water emulsions. Case Studies in Chemical and Environmental Engineering. 2023;*8*(September):100503. DOI: 10.1016/j.cscee.2023.100503

[73] Mohyaldinn ME, Alakbari FS, Nor A. N. A. BA, Hassan AM. Stability, rheological behavior, and PH responsiveness of CTAB/HCl acidic emulsion: Experimental investigation. ACS Omega. 2023;**8**(25):22428-22439. DOI: 10.1021/acsomega.2c08243

[74] Del Giudice F. A review of microfluidic devices for rheological characterisation. Micromachines. 2022;**13**(2):1-20. DOI: 10.3390/mi13020167

[75] Tadros T. Application of rheology for assessment and prediction of the long-term physical stability of emulsions. Advances in Colloid and Interface Science. 2004;**108-109**:227-258. DOI: 10.1016/j.cis.2003.10.025

[76] Hu YT, Ting Y, Hu JY, Hsieh SC. Techniques and methods to study functional characteristics of emulsion systems. Journal of Food and Drug Analysis. 2017;**25**(1):16-26. DOI: 10.1016/j.jfda.2016.10.021

[77] Pharmacopoeia E. The, A. 2.2.10. Viscosity - Rotating Viscometer Method. Strasbourg, France: European Pharmacopoeia; 2008. pp. 28-30. Available from: https://www.drugfuture.com/Pharmacopoeia/EP7/DATA/20210E.PDF

[78] Rohm H, Jaros D. Rheological Methods: Instrumentation; McSweeney, P.L.H., McNamara, J.P.B.T-E. of D.S. (Third; Academic Press: Oxford; 2022. pp. 530-536. DOI: 10.1016/B978-0-12-818766-1.00031-3

[79] Saasen A, Hodne H. Influence of vibrations on the rheological properties of drilling fluids and its consequence on solids control. Applied Rheology. 2016;**26**(2):1-6. DOI: 10.3933/APPLRHEOL-26-25349

[80] Rheology. In: Rig Worker. Available from: https://www.rigworker.com/fluids-3/rheology.html

[81] Azad MS, Patel V, Shah N, Sharma T, Trivedi JJ. Extensional rheological measurements of surfactant–polymer mixtures. ACS Omega. 2020;**5**(48):30787-30798. DOI: 10.1021/acsomega.0c00481

[82] Xu L, Liu X, Liu Z, Li P, Ding H, Gong H, et al. Extensional rheology of hydrophobically associating polyacrylamide solution used in chemical flooding: Effects of temperature, NaCl and surfactant. Chemical Engineering Science. 2023;**273**:118644. DOI: 10.1016/j.ces.2023.118644

[83] Sun Z, Pu W, Zhao R, Pang S. Study on the mechanism of W/O emulsion flooding to enhance oil recovery for heavy oil reservoir. Journal of Petroleum Science and Engineering. 2021;**2022**(209):109899. DOI: 10.1016/j.petrol.2021.109899

[84] Xu K, Zhu P, Colon T, Huh C, Balhoff M. A microfluidic investigation of the synergistic effect of nanoparticles and surfactants in macro-emulsion-based enhanced oil recovery. SPE Journal. 2017;**22**(2):459-469. DOI: 10.2118/179691-PA

[85] Karambeigi MS, Abbassi R, Roayaei E, Emadi MA. Emulsion flooding for enhanced oil recovery: Interactive optimization of phase behavior, microvisual and core-flood experiments. Journal of Industrial and Engineering Chemistry. 2015;**29**:382-391. DOI: 10.1016/j.jiec.2015.04.019

[86] Pryazhnikov M, Pryazhnikov A, Skorobogatova A, Minakov A, Ivleva Y. Microfluidic study of enhanced oil recovery during flooding with polyacrylamide polymer solutions. Micromachines. 2023;**14**(6):1-16. DOI: 10.3390/mi14061137

[87] Nilsson MA, Kulkarni R, Gerberich L, Hammond R, Singh R, Baumhoff E, et al. Effect of fluid rheology on enhanced oil recovery in a microfluidic sandstone device. Journal of. Non-Newtonian Fluid Mechanics. 2013;**202**:112-119. DOI: 10.1016/j.jnnfm.2013.09.011

[88] Pipe CJ, McKinley GH. Microfluidic Rheometry. Mechanics Research Communications. 2009;**36**(1):110-120. DOI: 10.1016/j.mechrescom.2008.08.009

[89] Ober TJ, Haward SJ, Pipe CJ, Soulages J, McKinley GH. Microfluidic extensional Rheometry using a hyperbolic contraction geometry. Rheologica Acta. 2013;**52**(6):529-546. DOI: 10.1007/s00397-013-0701-y

[90] Herbas JG, Group X, Wegner J, Hincapie RE, Ganzer L, Castillo JA, Del Mugizi HM, Exploration P. Comprehensive micromodel study to evaluate polymer EOR in unconsolidated sand reservoirs. SPE Middle East Oil Gas Conference. 2015;**1**:1-18

[91] Elhajjaji RR, Hincapie RE, Tahir M, Rock A, Wegner J, Ganzer L. Systematic study of viscoelastic properties during polymer-surfactant flooding in porous media. Soc. Pet. Eng. - SPE Russian Petroleum Technology Conference Exhibition. 2016;**2016**:2552-2569. DOI: 10.2118/181916-ms

[92] Hincapie RE, Duffy J, O'Grady C, Ganzer L. An approach to determine polymer viscoelasticity under flow through porous media by combining complementary rheological techniques. Soc. Pet. Eng. - SPE Asia Pacific Enhanced Oil Recovery Conference EORC. 2015;**2015**:1345-1361. DOI: 10.2118/174689-ms

[93] Jouenne S, Heurteux G. Correlation of mobility reduction of HPAM solutions at high velocity in porous medium with ex-situ measurements of elasticity. SPE Journal. 2020;**25**(1):465-480. DOI: 10.2118/198899-PA

[94] Abgrall P, Holloway J, Woodcock E, Lefeuvre Y, Adamska P, Meunier G. Fast determination of molecular weight using viscosity measurements using FLUIDICAM Rheo. Encyclopedia Article (Rheology on Chip). Toulouse, France: Mictrotrac Formulaction; 2019

Chapter 5

Application of Nanotechnology in the Petroleum Industry: A View from Rheology

Esteban Taborda, Yurany Villada, Lady J. Giraldo, Diana A. Estenoz, Camilo A. Franco and Farid B. Cortés

Abstract

The objective of this chapter is to demonstrate the use of rheology as a fundamental tool for fluid characterization in processes of the oil industry assisted by nanotechnology. In more detail, the chapter will present three important processes in the oil and gas industry: (i) increase the mobility and transport conditions of heavy and extra-heavy crude oil, (ii) improve the performance of drilling fluids, and (iii) improve polymer injection technology in enhanced recovery (EOR) processes, through the use of nanomaterials. To this end, steady-state rheometry and dynamic rheology are incorporated to understand the phenomena and relate the performance and microstructure of the fluids used. The content of the chapter presents interesting results of a disruptive technology of great importance to the energy industry in general. It will be of great technical contribution to the readers of the book.

Keywords: drilling fluids, heavy oil, nanotechnology, oil and gas industry, polymer flooding, rheology

1. Introduction

Rheology is the science that studies the flow response of fluids under certain forces applied that gives important information about the intrinsic properties and compositional microstructure of fluids [1]. It is known that the functional properties of fluids are dependent on their composition and structural disposition. These properties are favored under specific arrangements that promote a particular behavior for determined conditions of forces applied, pressure, and temperature, among others [1, 2]. Within the value chain of the Oil and Gas industry take great importance the flow of fluids inside rock/reservoirs and their transport through the pipes, where the understanding of its rheological behavior is critical [3–5]. In a general view of the value chain of the Oil and Gas industry are recognized the upstream, midstream, and downstream processes related to the exploration/production, transport/storage, and refining/use stages, respectively [6]. Upstream and midstream processes include well

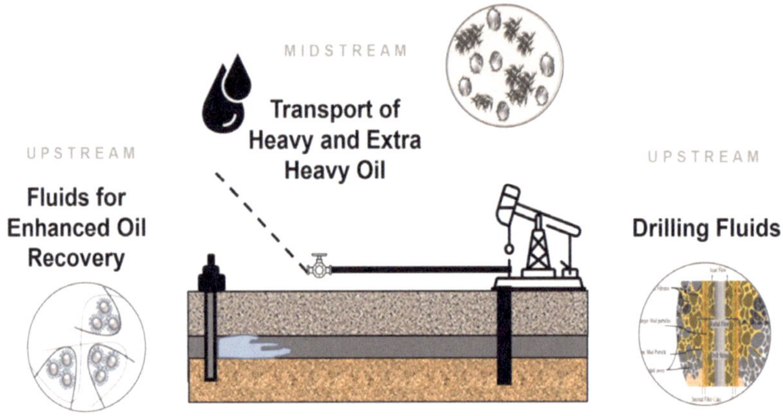

Figure 1.
Schematic representation of fluids used in stages of oil and gas operations. Images taken from Refs. [7–9].

drilling operations, chemical-enhanced oil recovery (CEOR) processes, and mobility and transport of crude oil (**Figure 1**) [6]. These processes have in common the proper fluid management with the design of water or mud-based drilling fluids, the selection of polymer or gels injection for enhanced oil recovery (EOR) application, and the respective flow considerations of crude oil inside the reservoir and the transport up to refining. For all of these fluids, the rheological behavior plays an important role in the adequate performance of each operation. Static and dynamic rheology allows an understanding of the flow behavior of several fluids applied in the oil industry [1]. The main tests for the characterization of these fluids include steady shear (flow curves) and oscillatory shear (amplitude, frequency, time, and temperature sweeps). The flow curves provide information about the flow behavior (Newtonian and non-Newtonian fluids) [1]. In particular, the fluids used in the described stages for oil extraction exhibit a non-Newtonian trend. Likewise, the frequency sweep indicates the elastic (G') and loss (G') modules used to evaluate the viscoelastic behavior of fluids. Temperature sweep is used to study the thermal stability of fluids from elastic and loss components, as well as the complex viscosity [1]. On the other hand, the time sweep is important to predict the gel formation in the fluids. This behavior is of vital importance to the drilling operation for suspending the cuttings under rest conditions. Rheological properties of these fluids can be impacted depending on the conditions of each field, reservoir fluids (brines and crude oil), temperature, and petrophysics properties of the reservoir, which can promote poor performance in drilling operations, low recovery factor with cEOR processes and low mobility and transport of heavy crude oil (HO) and extra-heavy crude oil (EHO).

The advances in nanotechnology in the last decades have shown the multiple applications and properties of nanomaterials in the different areas of science and Engineering [10, 11]. In this sense, the Oil and Gas industry has not been strange to this development with several applications from laboratory to field trial tests [12]. One of the main purposes of the nanomaterial's inclusion in fluids in the Oil and Gas industry has been focused on enhancing the rheological properties in upstream and midstream operations [13]. Some researchers have reported that nanoparticles can modify the viscoelastic network of fluids improving the rheological and thermal properties, among others.

2. Rheological behavior of heavy oils and influence of nanomaterials

Among the different types of reservoir fluids, heavy crude oil stands out. Currently, heavy and extra-heavy crude oil reserves represent 40–45% of total oil reserves worldwide [14, 15]; therefore, the need to continue extraction is current. The physical property that characterizes this class of hydrocarbons is their low API gravity ($\leq 20°$) and high viscosities [16–19]. This specific property is mainly due to the high content of asphaltenes and their tendency to self-associate, which in the presence of resins form a very stable viscoelastic network [20–26]; therefore, the mobility of these fluids is low, which forces the industry, in general, to constantly search for alternatives that allow improving the mobility of this type of fluids. However, there are limitations that range from the application of expensive technologies, the use of toxic substances, among other factors [27–33]. For this reason, in recent years, nanotechnology has emerged as a potential alternative capable of competing technically, economically, and environmentally with conventional processes [12, 34–39], based on the principle of reducing the viscosity of heavy crude oil due to the interaction of nanometric particles with the asphaltenes present in the crude [7, 26, 40–43]. Nanoparticles have characteristics such as high surface area-volume ratio, nanometric size, ease of synthesis, and, above all, high affinity for asphaltenes [42, 44–47], allowing their performance to be appropriate and producing a reduction in viscosity necessary to improve the mobility of heavy and extra-heavy crude oil.

Interesting results presented below demonstrate the performance of different types of nanoparticles in the rheological behavior of heavy crude oils.

2.1 Rheological measurements: steady-state and dynamic rheometry

Taborda et al. [5, 7, 40] evaluated different types of nanoparticles, from commercial to in-house synthesized, SiO_2 with acid, basic, and neutral surface, Al_2O_3, and Fe_3O_4 (more information in the supplementary material), in heavy crude oil matrices to evaluate their performance as viscosity-reducing agents, through steady-state and dynamic rheology measurements. The authors carried out the rheological measurement Kinexus pro Rheometer (Malvern, UK), using a plate-plate geometry with a 300 μm gap at 298 K. Each run was repeated three times. The oil and nanoparticles were mixed using an HP130915Q mixer from Thermo Scientific (Waltham, Massachusetts, USA), at 500 rpm for 30 min and a temperature of 298 K. They evaluated effects such as concentration of nanoparticles, particle size, chemical nature of the nanoparticles, and oil type, among others. The results are presented in **Figure 2**.

Figure 2(a) shows the rheological measurement of the HO matrix in the presence of S8 (8 nm diameter SiO_2 nanoparticles) at concentrations of 10, 100, 1000, and 10,000 mg/L. All nanoparticles evaluated generated viscosity reductions in heavy crude oil. The point of greatest viscosity reduction was found in the presence of 1000 mg/L of S8. The point of lowest viscosity reduction was found at a concentration of 10,000 mg/L. Despite interacting with asphaltenes, it is possible that it is due to increase in the packing factor of the nanoparticles, causing a reduction in the interaction energies with asphaltenes due to aggregation effects between the same nanoparticles.

All the flow curves presented in **Figure 2** exhibit a shear-thinning type behavior, typical of this class of fluids [49]. The degree of viscosity reduction determined by adding S8 nanoparticles varies from 3%, obtained at concentrations of 10 mg/L, to the

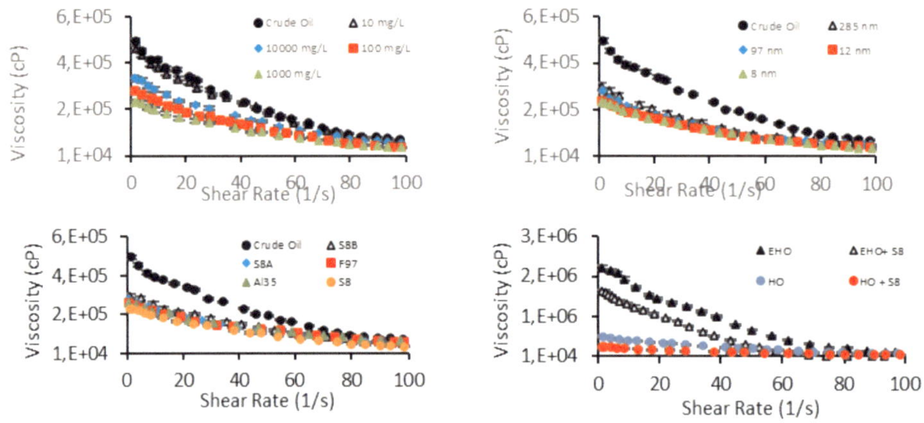

Figure 2.
Impact of nanoparticles on the flow curves of heavy crude oil. (a) effect of concentration, (b) effect of size, (c) effect of the chemical nature, (d) effect of the oil. Adapted from Taborda Acevedo [48].

point of greatest viscosity reduction, 45% obtained at 1000 mg/L. It is worth highlighting that as the shear rate increases, the performance in viscosity reduction is lower, due to the shear-thinning nature of the fluid, which causes changes in the microstructure of the fluid when subjected to greater stresses.

In **Figure 2(b)**, the effect of the nanoparticle size was evaluated, starting at 8 nm, up to 285 nm, using SiO_2 nanoparticles from different precursor sources, mixing commercial and in-house synthesized, all particles were added to 1000 mg/L. Similar to the previous results, the nanoparticles caused viscosity reductions where a clear trend of better performance is evident as the size of the nanoparticle is smaller. If the diameter is smaller, a greater number of particles are expected in the same amount of added mass compared to those with a larger diameter, this causes more points of contact between the nanoparticles and the crude oil, generating a greater degree of interaction between the asphaltenes and the nanoparticles.

To complete the evaluation of the properties of nanoparticles, the authors evaluated their different chemical natures. In **Figure 2(c)**, they evaluated S8 with a neutral (S8), acidic (S8A), and basic (S8B) surface, alumina (Al_2O_3), and magnetite (Fe_3O_4). The results demonstrated that the S8 nanoparticle with a neutral surface is the one with the best performance in viscosity reduction. The rheological evaluation of the performance of the S8 nanoparticles was completed by varying the type of crude oil between heavy and extra-heavy oil, the first one is heavy oil with 13°API, and the second is extra-heavy oil with 6.4°API; **Figure 2(d)** shows a very good performance of the nanoparticles, altering asphaltene colloidal structure producing reduction in viscosity. To understand the phenomenon more completely, dynamic rheology evaluations were carried out using the same nanoparticles in heavy crude oil systems.

Heavy crude oils can be categorized as viscoelastic liquids; that is, they have viscous behavior typical of pure liquids, but also elastic behavior typical of ideal solids. The measurements of the storage modules G' and loss G" are characterized by measuring the contribution of both ideal solid and pure liquid of each heavy crude oil. This is carried out through oscillometry measurements with sweep tests, both frequency and of strain. Before measuring the modules, the zone of linear viscoelasticity was identified in **Figure 3(a)** and **(b)**, which was found between 0.5 and 5% strain. The results presented show that G" is always greater than G', suggesting that viscous behavior predominates over elastic behavior in the process of fluid deformation and

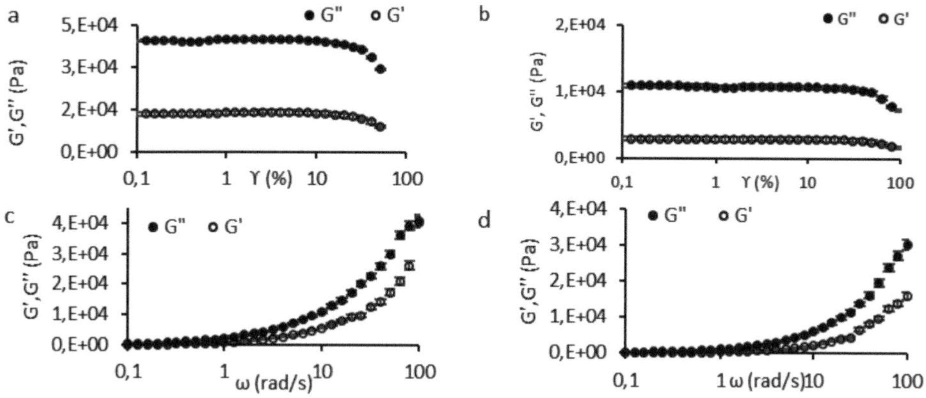

Figure 3.
Dynamic rheology measurements of heavy oil in the presence of S8 nanoparticles at 1000 mg/L and 298 K. (a) LVR of heavy oil, (b) LVR of heavy oil with S8 nanoparticles, (c) frequency sweep test of heavy oil, and (d) frequency sweep test of heavy oil with S8 nanoparticles. Adapted from Ref. [48].

flow. By adding 1000 mg/L of S8 nanoparticles, a clear reduction in the G' and G" modules is noted, altering the internal microstructure of the fluid, reducing the viscoelastic behavior of the material, favoring total mobility. **Figure 3(c)** and **(d)** show these results were obtained by executing the frequency sweep between 0.1 and 100 rad/s, at 298 K, at a fixed strain of 2% (within the linear viscoelasticity zone), this to eliminate noise generated by the natural movement of the sample in the rheometer.

The changes in the microstructure of the fluid are evidenced by the alterations in the viscoelastic moduli after the inclusion of the nanoparticles into the system; at a fixed frequency, the storage modulus G' decreases by 55%, and the loss modulus G" decreases by 64%, which explains the reduction in the viscosity of the heavy crude oil. This change in the microstructure of the fluid alters its rheological behavior in general. Rheological variables such as creep, elastic limit, and thixotropic behaviors are affected by the presence of nanoparticles as demonstrated by subsequent publications [50–52]. These rheological changes favor the mobility of the fluids, evaluated at different conditions, and generate a positive impact on the transportation and mobility processes of heavy and extra-heavy crude oil.

The reduction in viscosity produced by the presence of nanoparticles is striking; for this reason, a new rheological model was proposed, seeking to understand phenomenologically how the particles work and to be able to predict viscosities at different concentrations of nanoparticles in the crude oil. The model is presented below.

2.2 HB-PR modified model

The model presented by Taborda et al. [53] allows us to relate the pseudoplastic behavior of heavy crude oil with the inclusion of the Herschel-Bulkley model, and the rheological behavior of suspensions in the presence of nanoparticles with a mathematical modification to the Pal and Rhodes model. This model allows the experimental data to be correlated correctly. The modified Pal and Rhodes model is a mathematical approach presented in Ref. [54] as shown in Eq. (1):

$$n_r = (1 + K_o\phi)^{-V} \tag{1}$$

Where n_r is the relative viscosity, ϕ is the volume fraction of nanoparticles, V is the "shape factor" of dispersed particles, and K_o is the solvation constant, for spherical particles takes a value of 2.5.

One of the most widely used rheological models to describe the flow behavior of pseudoplastic liquids is the Herschel-Bulkley (H-B) model. The development of the model is presented in Refs. [55, 56]. This model was selected to evaluate mixtures of heavy crude oil originating from the Gulf of Mexico and called Mb, and S8 nanoparticles. The H-B model is described as follows:

$$\mu = K_H\left(\gamma^{(n_H-1)}\right) + \mu_{\infty,\gamma} \tag{2}$$

Combining Eqs. (1) and (2) leads to the Herschel-Bulkley-Modified Pal and Rhodes model (HB-MPR) as shown in Eq. (3):

$$\mu = \left[K_H\left(\gamma^{(n_H-1)}\right) + \mu_{\infty,\gamma}\right](1 + K_o\phi)^{-V} \tag{3}$$

The model enables a full understanding impact of nanoparticles on the viscosity of HO and EHO when submitted to a certain shear. **Figure 4** presents the evaluation of the model in another heavy crude oil in the presence of S8 nanoparticles at different concentrations. **Table 1** presents the parameters of the proposed model. A clear trend is noted in all the parameters that follow the decrease in viscosity.

The impacts that these findings have generated can be evidenced by multiple investigations developed subsequently, where mobility processes in the reservoir, surface transportation conditions, and durability of treatments over time, among others, were evaluated [5, 51, 52].

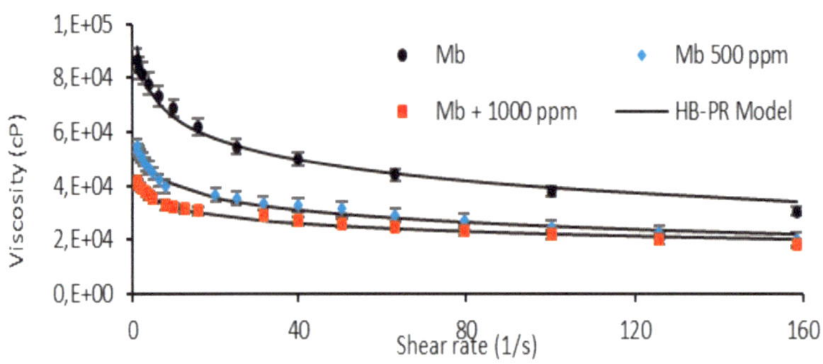

Figure 4.
Extra-heavy oil experimental and HB-PB proposed model plots. Figure taken from Ref. [53].

Np's type	Np's concentration (mg/L)	$\mu_{\infty,\gamma}$(cP)	$K_H \times 10^5$	n_H	$K_o \times 10^3$	V	RSME%
	0	512	0.99	0.80	NA	NA	7.9
S8	500	388	0.645	0.83	39.8	0.214	4.6
	1000	212	0.557	0.85	43.1	0.247	4.9

Table 1.
HB-PR proposed model parameters.

3. Rheological properties of drilling fluids

Drilling operation is a common and important process in the exploitation of hydrocarbons. The design and formulation of a proper drilling fluid are essential for a successful operation. The design and composition of these fluids are based on the characteristics of the geological formations and the available additives. Likewise, the fluid should fulfill physicochemical, economic, and environmental requirements [57]. Drilling muds serve several functions: provide hydrostatic pressure to protect invasion of the wellbore by formation fluids; promote sealing of porous formation; prevent damage to the formation; clean, cool, and lubricate the bit and other components; allow the suspension and mobility of drilling cuttings from subsurface to surface; and ensure well stability [58]. Two types of drilling fluids are conventionally used: water-based (WBMs) and oil-based (OBMs). The OBMs have high performance for the penetration rate, excellent filtration control, high lubricity, thermal stability, high salt tolerance, and good ability to transport cuttings, and they promote a decrease in the swelling of the clay formation and the stabilization of the well [59]. However, OBMs have disadvantages related to the environmental impact, cost, and negative effects on well cementing [60]. In contrast, WBMs have a low environmental impact but present some disadvantages in terms of the shale inhibition, lubricity, and thermal stability.

Due to the advancement of nanotechnology, the implementation of nanoparticles in the oil and gas industry has been proposed. Recently, the use of nanoparticles has received significant attention due to their effects on the rheological and filtration properties of water-based fluids (WBMs), the stability of the shale, the well strengthening, and the stability of fluids under hard conditions (high temperature and high pressure) [61]. Inorganic nanoparticles such as silica graphene oxide, nanographite, copper oxide, zinc oxide, tin oxide, aluminum oxide, iron oxide, carbon nanotubes, and composite nanoparticles such as polyacrylamide/polyethylene glycol, attapulgite/zinc oxide, polyacrylamide/nanoclays, carboxymethylcellulose/polystyrene, titanium oxide/polyacrylamide, alumina/polyacrylamide have been studied. Organic nanoparticles, such as nanocrystalline nanofibrillar cellulose and chitin nanocrystals, have also been probed [62].

Villada et al. [63] studied the use of cellulose nanofibrils as a replacement of xanthan gum in drilling fluids for shale formation. Two types of nanoparticles were incorporated. Bleached (B-CNF) and unbleached (L-CNF) cellulose nanofibrils, mainly differentiated by their lignin content, were probed. The authors employed rheometric assays to analyze the effect of nanoparticles on the viscosity of WBMs. **Figure 5** presents the flow curves corresponding to the WBMs designed. The results showed that all WBMs have a shear-thinning behavior. Likewise, the rheological properties increase with nanoparticle concentration. However, this effect is more noticeable for nanoparticles containing lower lignin amounts (BCNF). On the other hand, several rheological models were implemented to determine theoretically the rheological behavior. Particularly, the Sisko model predicted the rheological behavior of studied WBMs (Eq. (4)). The authors argued that the structural differences in dimension, shape, surfaces and thermal characteristics, rheological properties, and lignin content between B-CNF and L-CNF produced several effects on the rheological and filtration properties of WBMs. In this sense, changes in the viscosity, filtrate volumes, thickness, permeability, and filtration rate of the filter cake, and in the structure of the fluid were observed.

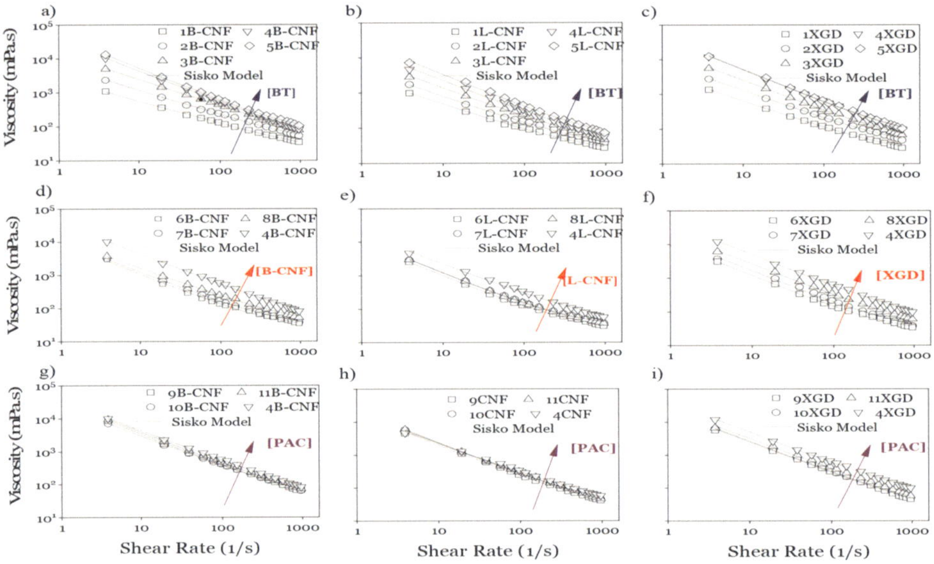

Figure 5.
Flow curves for WBMs designed. The steady shear analyses were performed at 25°C and 1 atm. The figure was adapted from Villada et al. [63].

Sisko model

$$\eta = \eta_{\infty} + k\dot{\gamma}^{n-1} \tag{4}$$

where $\eta\infty$ (mPa.s) is the infinite shear viscosity, k is the consistency index (mPa.s), $\dot{\gamma}$ (1/s) is the shear rate, and n is the flow behavior index.

The use of di-carboxylic acid cellulose nanofibril (DCA-CNF) as an additive in water-based drilling fluids (WBMs) applied to shale formations was proposed by Villada et al. [64]. The authors used a DV3TRV viscometer (Brookfield, USA) with a cone-plate geometry (CP-51Z with 2.4 diameter and 1.56 angle) to determine the steady-shear rheological behavior of DCA-CNF aqueous suspensions and WBMs at 25° C. The flow curves were performed three times. The viscoelastic properties of WBMs were also investigated using oscillatory rheology. The tests were performed in a Haake RheoStress RS80 rheometer (Haake Instruments Inc., Paramus, NJ, USA) with a cone-plate geometry (60 mm diameter, 1° angle). Silicone oil was added to the edge of the WBMs to prevent evaporation of the WBMs.

Frequency sweep tests were carried out from 0.01 to 10 Hz within the linear viscoelastic range at strains of 0.02 and room temperature. The test was repeated two times. Before each assay, the drilling fluids were stirred at 5000 rpm for 3 min. On the other hand, time and temperature sweeps were conducted. The time sweep was performed at constant stress (γ =0.02), temperature (25°C), and frequency (1.00 Hz). The evaluated time range was from 0 to 3600 s. To perform this test, any shear history in the fluid must be removed. Thus, a rotational shear test was carried out at a shear rate of 1000 1/s for 180 s. The temperature sweep was evaluated from 10 to 80°C, with a heating rate of 1.55°C/min, strain (0.02), and frequency (1.0 Hz). Before temperature sweep, an amplitude sweep was performed at the higher evaluated temperature (80°C).

The authors observed a shear-thinning behavior associated with the changes in viscosity in the presence of nanoparticles. Likewise, the rheological properties of WBMs increased with the concentration of nanoparticles. Furthermore, the Sisko model was fitted very well with experimental data. The viscoelastic properties of WBMs are presented in **Figures 6–9**. The frequency sweeps for studied WBMs are presented in **Figure 6**. The mechanical spectra show that fluids have a viscoelastic behavior. In addition, the elastic modulus (G') is higher than the viscous modulus (G"), which is an indication of dominant elastic and gel properties. This result could be related to the existence of fluids with a high degree of structuring favored by physical crosslinks between polymer chains and their interactions. Besides, the fluids exhibit a shear-thinning behavior, and the complex viscosity decreases with the shear rate. This behavior is in concordance with the results obtained from steady shear analyses.

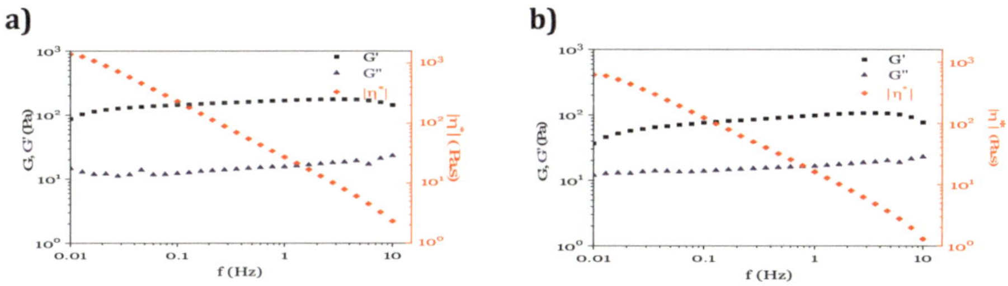

Figure 6.
Mechanical spectrum of WBMs: (a) fluid 4DCA-CNF and (b) fluid 4XGD. The figure was adapted from Villada et al. [64].

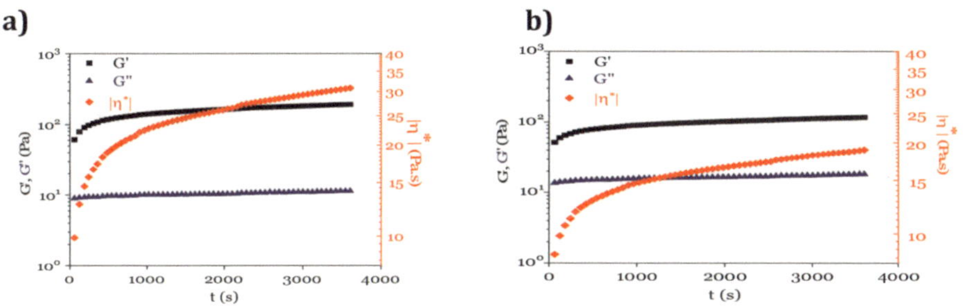

Figure 7.
Time sweep: (a) fluid 4DCA-CNF and (b) fluid 4XGD.

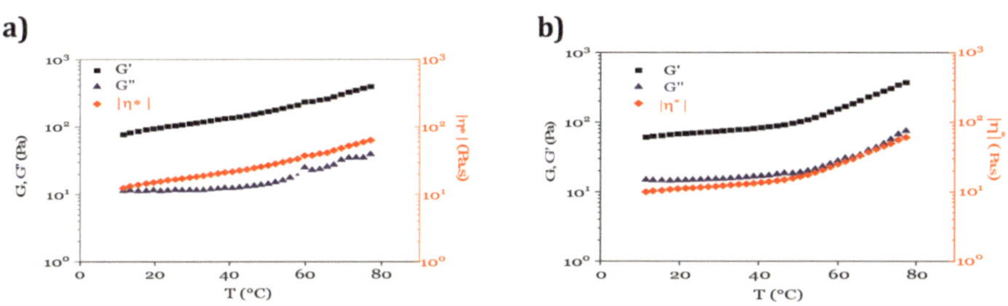

Figure 8.
Temperature sweep: (a) fluid 4DCA-CNF, and (b) fluid 4XGD.

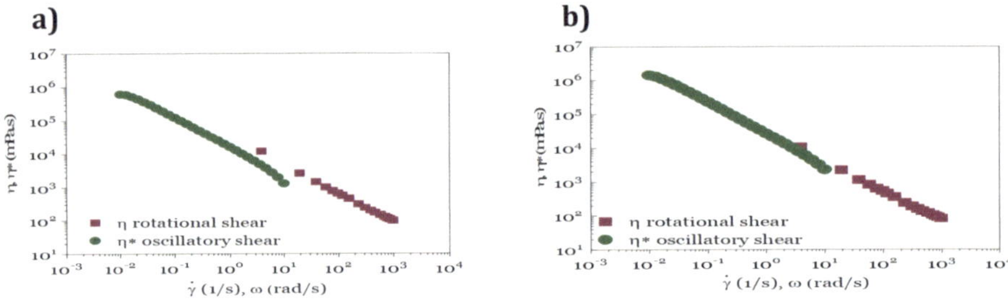

Figure 9.
Comparison between the viscosity of rotational shear and oscillatory shear (Cox-Merz rule) for the fluids studied: (a) fluid 4DCA-CNF and (b) fluid 4XGD.

Figure 7 depicts the variation of G' and G" as a function time for the fluids 4DCA-CNF, and 4XGD. The results show values of G' > G" for both fluids, which indicates again more elastic than viscous properties. The fluids exhibit a gel structure with stable formation at times greater than 600 s (maximum value of G') and 1.00 Hz. The estimated time follows the results reported by Bui et al. [65] who associated the structural stability with the constant value of G'. On the other hand, the complex viscosity increases with time. This behavior is attributed to the interaction of the additives present in the WBMs.

Overall, the DCA-CNF fluid shows higher viscoelastic parameters (G',η*), suggesting that some electrostatic interactions between the bentonite, DCA-CNF, and the polyanionic cellulose are more favorable. Compared to the two fluids, there is a greater difference in the values of the elastic than the viscous component.

Figure 8 shows the variation in the elastic and viscous moduli and complex viscosity with temperature. It is observed in both fluids that G' > G", which indicates a greater elastic behavior. This behavior is more noticeable for the DCA-CNF fluid. The fluid presents two linear increases of G' and η* around 56 and 80°C. In addition, G" exhibits a constant value from 10 to 40°C and a linear increase from 40 to 80°C, **Figure 8(a)**. The increase in G' from 10 to 40°C suggests the formation of a more rigid structure in the fluid whereas G" remained constant. The modulus increases with temperature, indicating a similar trend from 56 to 80°C. On the other hand, the 4XGD fluid presents a linear increase in G' and η* from 10 to 48°C, and in the temperature range of 48–80°C. Besides, G" exhibits a constant value and pronounced linear increase, **Figure 8(b)**.

The comparison between the viscosity of rotational shear and oscillatory shear using the Cox-Merz Rule (Eq. (5)) for the studied WBMs is presented in **Figure 9**. The results indicate that the WBMs follow the Cox-Merz rule due to the similarity in both viscosities. Particularly, the fluid 4DCA-CNF shows a complete overlap, whereas, for the fluid 4XGD, a slight deviation is observed. This behavior is related to the experimental variations. The application of the Cox-Merz Rule allows the evaluation of the behavior of drilling fluids under low and high shear rates.

$$\eta^*(\omega) = \eta(\dot{\gamma})|_{\omega=\dot{\gamma}} \tag{5}$$

The incorporation of calcium carbonate particles ($CaCO_3$) as an additive to improve the rheological, filtration, thermal, and structural properties of drilling fluids was studied by Villada et al. [66]. $CaCO_3$ particles were synthesized by the precipitation method and characterized in terms of morphological, thermal, and structural properties. WBMs were designed containing three different viscosifier additives such as xanthan gum, (XGD), guar gum (GG), or corona gum (GEC). The authors evaluated the effect of particles on the rheological properties through steady shear analyses. The assays were performed with a Brookfield DV3TRV (cone/plate) viscometer (Brookfield, USA). The cone-plate CP 51Z configuration was used with a shear rate range of 3.84–960 1/s at 25°C. Before each viscosity measurement, the samples were pre-sheared at 960 s^{-1} for 30 s to erase any shear history. The assays were performed in triplicate. Villada et al. [66] adjusted two empirical models: Sisko (Eq. (4)) and Herschel-Bulkley (Eq. (6)).

$$\tau = \tau_0 + k'\gamma^{n'} \tag{6}$$

where τ_0(Pa.s) is the yield stress, *k*' is the consistency index (mPa.s), $\dot{\gamma}$ (1/s) is the shear rate, and *n*' is the flow behavior index.

The results showed that fluids containing XGD and $CaCO_3$ particles exhibited significant improvements in their functional properties. These results were associated with the surface interactions between bentonite with polymers and bentonite with particles.

Compared to the fluids formulated with GG and ECG, the use of XGD in the WBMs resulted in a different rheological behavior that could related to the formation of a more effective coating on the surface of Na-Bentonite platelets. Overall, the authors argued that the complex microstructure of multicomponent systems (clay, $CaCO_3$ particles, and polymers) is mainly determined by the association mode of clay, the charge of $CaCO_3$ particles, and the nature of polymer components that modify the rheological behavior of WBMs. Moreover, the rheological behavior of WBMs was simulated using different empirical rheological models to provide a physical interpretation of flow behavior.

4. Viscoelastic properties of injection fluids for cEOR processes

During the productivity life of hydrocarbon fields, enhanced oil recovery processes play an important role in increasing well productivity. The injection of fluids in injection-producer patterns is one of the main methodologies for improving oil recovery. Polymers, surfactants, and gels are common chemical agents used for fluid injection based on capillary and viscous properties that can be impacted by them in interaction with the reservoir fluids. Polymer/gel solution injection can have two purposes: (a) for mobility control of displacing fluids for oil displacement increasing oil recovery and (b) conformance control operations to improve sweep efficiency of CEOR process blocking high permeability zones to allow fluids flow through non-invaded zones. In both cases, the rheological properties of fluids (viscosity and viscoelastic properties) have crucial importance in the design of polymeric/gel fluids. However, due to conditions of high temperature, ion content of brines, the presence of oxygen and hydrogen sulfide, petrophysical properties of the reservoir, and adsorption onto the rock, the performance of polymer/gel solutions can be weakened, affecting their stability, and decreasing their effectiveness. Thereby, nanotechnology emerges as a promissory alternative to enhance the rheological properties and diminish the negative effects. Nanomaterials present exceptional properties of size and surface energy promoting physicochemical interactions with certain chemicals, which can produce synergistic effects improving thermal stability and salinity tolerance, increasing the gel strength and storage modulus [12, 67–69].

Polymer flooding as a chemical-enhanced oil recovery technique is designed for mobility control of displacing fluids to modify the mobility ratio between displacing and displaced fluids in the reservoir leading to higher oil recovery. Currently, the hydrocarbon production of the majority of the mature fields in the world is based on water flooding followed by tertiary recovery or EOR methodologies. The cEOR techniques as part of EOR methodologies included polymer flooding as a substantial modification of the water flooding stage by polymeric additives on the fluid injection carrier (water/brine). The addition of polymeric agents contributes to the viscous properties of injection fluid that favor an improvement in oil sweep efficiency in the reservoir. The increase of the viscous effects on injection fluids decreases its fingering path trend in the reservoir leading to a better sweep and higher oil displacement. Nanotechnology approaches for polymer flooding present an important impact in terms of enhancing the polymer properties such as thermal stability and salinity tolerance to improve the higher performance of polymeric solutions on reservoir. Several authors have studied the effect of different chemical natures of nanoparticles such as Al_2O_3, SiO_2, Ni_2O_3, ZnO, MgO, Fe_3O_4, TiO_2, ZrO_2 on the rheological behavior of polymer-based fluids [70].

Al-Anssari et al. [70] evaluated the effect of zirconium nanoparticles mixed with polyacrylamide (PAM) polymer on viscosity measurements under different temperatures (from 25 to 75°C) and the presence of high salinity (1%wt NaCl). The viscosity measurements were performed using a rheometer (discovery HR-3 hybrid) with the respective temperature controller. The authors found that nanoparticle inclusion increases the viscosity of polymeric solution even at harsh conditions of temperature or salinity content. They concluded that nanoparticles can act as a crosslinker agent with the polymer chains forming a rigid polymeric network, developing a new hybrid structure by nanoparticle integration with enhanced rheological properties [70]. On other hand, Giraldo et al. [71] studied the effect of silica nanoparticles on the thermal stability and rheological behavior of HPAM-polymer-based solutions. The rheological measurements were conducted in a kinexus pro rotational rheometer using the concentric cylinder geometry, equipped with a Peltier cylinder cartridge for control of temperature. The authors found that nanoparticles improve the thermal stability of polymer at elevated temperatures of 70°C for 7 and 14 days. **Figure 10** presents the results obtained with a decrease in the degree of polymer degradation with values of 4.5 and 16% for 3000 and 500 mg/L of SiO_2 nanoparticles, respectively. On the contrary, the polymeric solution in the absence of

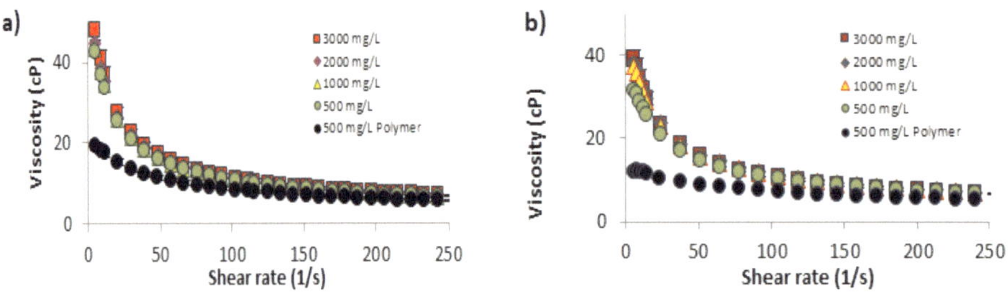

Figure 10.
Rheological behavior and degree degradation of 500 mg/L polymer solutions in the presence of different dosages of SiO2 nanoparticles after degradation at 70°C under an inert atmosphere for (a) 7 and (b) 14 days. Figures adapted from Giraldo et al. [71].

nanoparticles presents a degradation degree of 65% at 7 days. It is important to note that all polymer-based fluids evaluated exhibit a shear-thinning behavior which is a representative behavior of this type of fluid.

More recently, Zapata et al. [72] assessed silica nanoparticle amine functionalized to inhibit the chemical degradation by ion presence on HPAM/Brines solutions. The authors evaluated the effect of mono-, di-, and trivalent cations' presence in brine on the viscous properties of polymeric solutions. The results suggest a direct correlation between the ion's valence and the percentage of degradation. Polymer preparation was conducted following API 63 standard specifications. Rheological measurements were performed in a Kinexus pro rheometer with a concenter cylinder geometry. Fluids evaluated present a shear-thinning behavior. The authors highlight the effect of functionalized nanomaterials in addition to polymeric solutions improving the viscous properties by up to 16% in comparison with systems without nanomaterials. Besides, the authors include molecular simulation analysis to describe the phenomenological interactions that occurred to explain the effect on the viscous properties.

Conformance control operations are deployed to improve the oil-swept efficiency in productive zones in the reservoir. The goal of these operations is to obtain additional oil trapped in unswept areas with low permeability that, in usual paths of EOR fluid injections, are not able to mobilize. Traditional gels used are formed by polymers (of high molecular weight) mixed with organic or metallic agents that act as crosslinkers. Additives like partially hydrolyzed polyacrylamide (HPAM), sulfonated polyacrylamide (SPAM), polyacrylamide (PAM), xanthan gum, guar gum, polyvinyl alcohol (PVA), and crosslinkers like chromium (III), aluminum (III), zirconium (III), polyethyleneimine, formaldehyde, among other are widely reported in the literature [68, 73]. Several authors have evaluated the impact of nanoparticle addition in the preparation of gel for conformance. Obino et al. [68] and Fathima et al. [67] present a comprehensive review of different types of nanomaterials used for this purpose, highlighting their performance for thermal stability and gel strength. The use of nanomaterials of different chemical nature has been reported. Metal and metal oxide-based (like oxides of aluminum, titanium, zinc, iron, nickel, and magnesium, among others), organic (graphene and its derivatives, nanotubes, among others), silica nanomaterial and its surface modifications, among others [67, 68].

Thoniyot et al. [74] make an approximation of nanoparticle-hydrogel composite interaction for gel formation. They argued that nanoparticles can act as crosslinkers according to their specific surface properties interact with the functional groups of polymer to form strong networks [74]. In the same way, other authors that include nanoparticles in combination with gel systems also refer to the crosslinker effect of nanomaterials that leads to a stronger viscoelastic network in the fluid [75–78]. Nonetheless, for conformance applications, some authors also evaluate the viscoelastic behavior of gel by qualitative bottle test through Sydansk code which gives qualitative information about gelation time and gel strength [79]. Sydansk code describes with alphabetic letters the flow and consistency of gel formed, where A, B, C, D, and E are assigned according to the flow of gel from no detectable gel formation to barely flowing gel, while F, G, and H refer to non-flowing gels, and I and H for rigid gel appearance [79]. Nonetheless, it is important to note that analysis of static and dynamic rheological behavior as curves of apparent viscosity versus shear rate and oscillatory measurement allows a deeper understanding of the microstructure of fluids and its considerations of solid-like or liquid-like behavior.

Perez et al. [80] studied the effect of SiO_2, MgO, Cr_2O_3, and Al_2O_3 nanoparticles on the rheological properties of acrylamide sodium acrylate copolymer/Cr(III) acetate gel for conformance control application in reservoirs. The authors highlight the good results of nanoparticle addition in terms of improved gel strength assessed by static and dynamic rheological evaluation and syneresis test. Initially, the polymer solution was prepared, and then, the crosslinker was (Cr(III) acetate) added drop by drop in a ratio of 40:1 under constant stirring. The nanoparticles were added in the final step after the crosslinker addition under constant mechanical stirring. The rheological analysis was conducted by viscosity and oscillatory measurements for the understanding of gel microstructure with the storage G’ and loss modulus G” in the presence and absence of nanoparticles. A Kinexus pro-rotational rheometer was used for the rheological evaluation and equipped with a Peltier plate and cylinder cartridges for temperature control. For viscosity measurement, a Couette geometry was used and cone-plate geometry for G and G” with an oscillation frequency from 1 to 100 Hz and 5% of strain. **Figure 11** presents the curves of apparent viscosity *vs* time to the gelation time of gels prepared at 4000 and 8000 mg/L of polymer. An increase in the viscosity of the polymer solutions is observed due to the addition of the nanoparticles. Furthermore, a noticeable inflection point between 750 and 1500 s is observed, indicating the gel formation. This behavior is associated with tridimensional network formation due to the crosslinking effect of nanoparticles in the gel structure. As the concentration of polymer increases, the gelation time is reduced due to the great availability of the carboxylic group of the polymer chain and crosslinker molecules interacting to create bonds [80]. In another similar work reported by Perez et al. [81] with a gel system formed by hydrolyzed polyacrylamide/resorcinol/formaldehyde also were assessed SiO_2, MgO, Cr_2O_3, and Al_2O_3 nanoparticles following the methodology previously mentioned. The authors found similar results in terms of a higher gel strength by nanoparticle addition, with the increase of storage modulus G’ as evidence of the predominant solid-like behavior of gel. Additionally, an increase in nanoparticle concentration produces higher values of G, observing the best performance with Al_2O_3 nanoparticles, followed by MgO and Cr_2O_3, while SiO_2 nanoparticles do not show significant influence, as presented in **Figure 12**.

Nevertheless, **Figure 13** presents the frequency sweep of gel prepared with a dosage of 100 mg/L of nanoparticles at 4000 and 8000 mg/L of polymer concentration in gel systems. The authors show that in the range of frequency assessed, the modulus G” is slightly higher than the G’ modulus for gels prepared in the absence of nanoparticles, which makes sense due to the predominant liquid-like behavior over

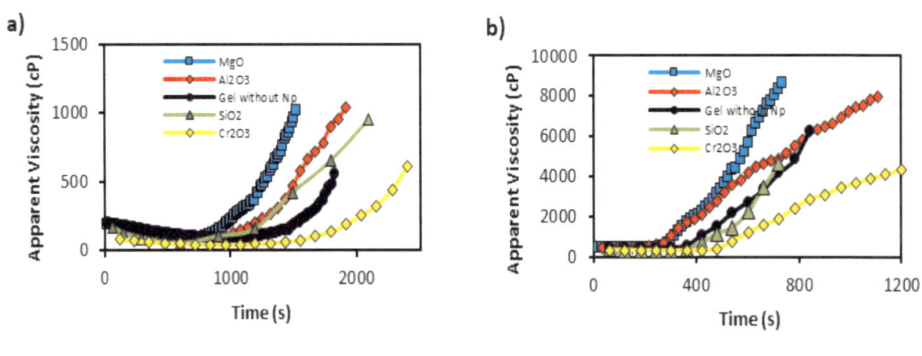

Figure 11.
The gelation time of gel prepared with acrylamide sodium acrylate copolymer/Cr(III) acetate in the absence and presence of nanoparticles at (a) 4000 mg/L and (b) 8000 mg/L of polymer concentration and 70°C. Figures adapted from Perez et al. [81].

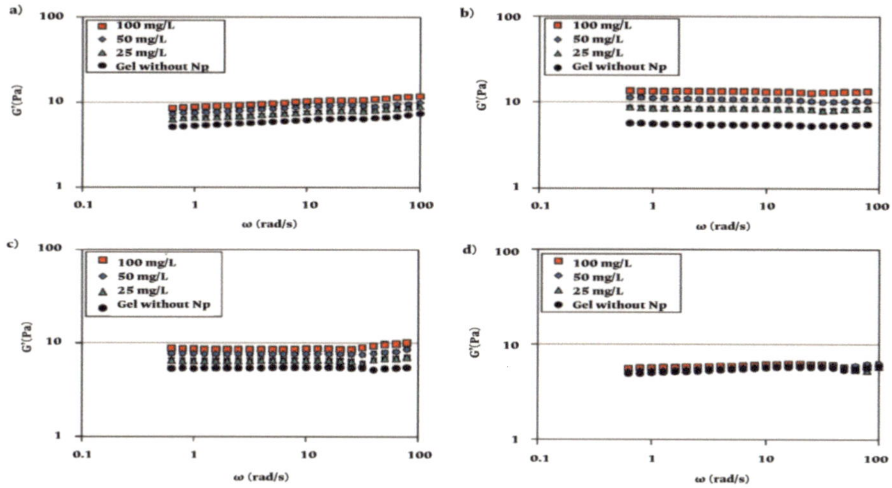

Figure 12.
Storage modulus for gel system (0.3% HPAM, 0.06% resorcinol, and 0.4% formaldehyde) with the addition of nanoparticles at different dosages (a) MgO, (b) Al_2O_3, (c) Cr_2O_3, and (d) SiO_2 nanoparticles at 25°C. Figures adapted from Perez et al. [80].

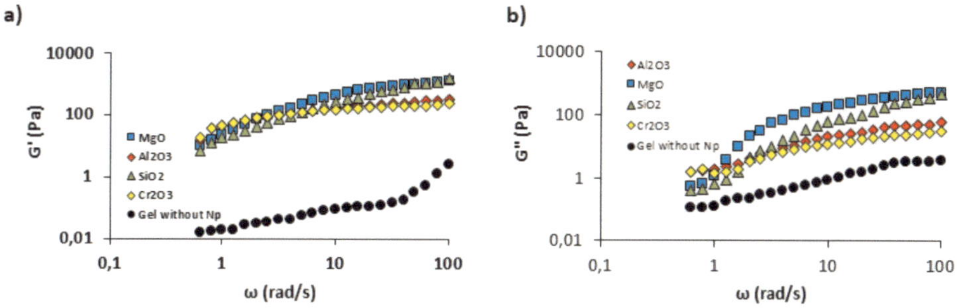

Figure 13.
Storage and loss modulus of acrylamide sodium acrylate copolymer/Cr(III) acetate gel with the dosage of 100 mg/L of nanoparticles and polymer concentration of 4000 mg/L. measured at 70°C. Figures adapted from Pérez-Robles et al. [80].

the solid-like behavior. However, when the nanoparticles were added, the viscous response was diminished. The authors argue the results of interactions between nanoparticles, polymer, and crosslinker based on the formation of nucleation points by nanoparticles which decrease the entanglement [81]. A general trend is observed in **Figure 13** for 4000 and 8000 mg/L of polymer in gel systems, where G is greater than G" and elastic modulus increases with the presence of nanoparticles based on the strengthening of the tridimensional network. At low polymer concentration (4000 mg/L) is marked the impact of nanoparticles with an improved gel structure, observing the best performance with Al_2O_3 y Cr_2O_3 nanoparticles. Based on this result, the author suggests that the inclusion of nanoparticles could also be used to reduce the amount of polymer used in the gel systems maintaining similar properties, which is an important issue for the cost/benefit relationship.

Rheological analysis has a critical importance for the adequate selection of gel/nanoparticles system for conformance applications where a high gel strength is desired, but also a delayed gel formation could be favorable for injection into the target zone in the reservoir, being these properties contributed by nanoparticles addition where can be solid-like behavior dominate over liquid-like [81].

Finally, the gel strength evidenced by the rheological behavior in the presence and absence of nanoparticles is also corroborated with the syneresis test with similar results of gelation time. The displacement test results shows an evident increase of oil recovered then of plugging of high permeability zone obtaining in the low and medium permeability slim tubes 55.9 and 45% of incremental oil recovery volume, respectively.

5. Conclusions

The inclusion of nanotechnology in the petroleum industry generates technical benefits that enhance industrial operations necessary to facilitate the exploitation and management of hydrocarbons, from the preparation of wells in drilling processes to improving the techniques that allow increasing the productivity of the fields, to enhance transportation and mobility once the fluids reach the surface. To understand how nanotechnology works, rheology appears as a fundamental science that helps to explain the phenomenology of processes enhanced with nanotechnology. Interesting results include the improvement of drilling fluids, which allows optimization of their formulation and subsequent performance of the technique, reducing the viscosity of heavy and extra-heavy crude oil, optimizing gel addition processes to allow the selective flow of hydrocarbons in the reservoir, reducing the degradation of polymers in EOR technologies that allow greater hydrocarbon sweeps.

Acknowledgements

This study was financed by the Francisco José de Caldas Fund, MINCIENCIAS and the National Hydrocarbons Agency (ANH) through contract No. 112721-282-2023 (Project 1118-1035-9300) with the National University of Colombia - Medellín Headquarters and Parex Resources Colombia AG. Sucursal.

Author details

Esteban Taborda[1*], Yurany Villada[1], Lady J. Giraldo[1], Diana A. Estenoz[2], Camilo A. Franco[1] and Farid B. Cortés[1]

1 Universidad Nacional de Colombia, Medellín, Colombia

2 Universidad Nacional del Litoral, Santa Fé, Argentina

*Address all correspondence to: eatabord@unal.edu.co

References

[1] Barnes HA. A Handbook of Elementary Rheology. Cardiff, Wales: University of Wales, Institute of Non-Newtonian Fluid Mechanics; 2000

[2] Chen DT, Wen Q, Janmey PA, Crocker JC, Yodh AG. Rheology of soft materials. Condensed Matter Physics. 2010;**1**:301-322

[3] Rousseau D et al. Rheology and transport in porous media of new water shutoff/conformance control microgels. In: SPE International Conference on Oilfield Chemistry? The Woodlands, Texas. 2005. SPE-93254-MS: SPE

[4] Sorbie K, Clifford P, Jones E. The rheology of pseudoplastic fluids in porous media using network modeling. Journal of Colloid Interface Science. 1989;**130**(2):508-534

[5] Taborda EA, Franco CA, Lopera SH, Alvarado V, Cortés FB. Effect of nanoparticles/nanofluids on the rheology of heavy crude oil and its mobility on porous media at reservoir conditions. Fuel. 2016;**184**:222-232

[6] Patidar AK, Agarwal U, Das U, Choudhury T. Understanding the oil and gas sector and its processes: Upstream, downstream. In: Understanding Data Analytics and Predictive Modelling in the Oil and Gas Industry. Florida, USA: CRC Press; 2024. pp. 1-20

[7] Taborda EA, Alvarado V, Franco CA, Cortés FB. Rheological demonstration of alteration in the heavy crude oil fluid structure upon addition of nanoparticles. Fuel. 2017;**189**:322-333

[8] Magzoub MI, Salehi S, Hussein IA, Nasser MS. Investigation of filter cake evolution in carbonate formation using polymer-based drilling fluid. ACS Omega. 2021;**6**(9):6231-6239

[9] Llanos SN, Giraldo LJ, Santamaria O, Franco CA, Cortés FB. Effect of sodium oleate surfactant concentration grafted onto SiO2 nanoparticles in polymer flooding processes. ASC Omega. 2018; **3**(12):18673-18684

[10] Mobasser S, Firoozi AA. Review of nanotechnology applications in science and engineering. Journal of Civil Engineering and Urbanism. 2016;**6**(4): 84-93

[11] Nasrollahzadeh M, Sajadi SM, Sajjadi M, Issaabadi Z. Applications of nanotechnology in daily life. Interface Science Technology. 2019;**28**:113-143

[12] Franco CA, Franco CA, Zabala RD, Bahamón I, Forero A, Cortés FB. Field applications of nanotechnology in the oil and gas industry: Recent advances and perspectives. Energy & Fuels. 2021; **35**(23):19266-19287

[13] Boul PJ, Ajayan PM. Nanotechnology research and development in upstream oil and gas. Energy Technology. 2020; **8**(1):1901216

[14] Guo Z, Sun S, Wang Y, Ni J, Qian X. Impact of new energy vehicle development on china's crude oil imports: An empirical analysis. World Electric Vehicle Journal. 2023;**14**(2):46

[15] Wang Z, Li S, Jin Z, Li Z, Liu Q, Zhang K. Oil and gas pathway to net-zero: Review and outlook. Energy Strategy Reviews. 2023;**45**:101048

[16] Chew KJ. The future of oil: Unconventional fossil fuels. Philosophical Transactions of the Royal

Society of London A: Mathematical, Physical and Engineering Sciences. 2014; **372**(2006):20120324

[17] Meyer RF, Attanasi ED. Heavy oil and natural bitumen-strategic petroleum resources. World. 2003;**434**:650-657

[18] Rana MS, Samano V, Ancheyta J, Diaz J. A review of recent advances on process technologies for upgrading of heavy oils and residua. Fuel. 2007;**86**(9): 1216-1231

[19] Shah A, Fishwick R, Wood J, Leeke G, Rigby S, Greaves M. A review of novel techniques for heavy oil and bitumen extraction and upgrading. Energy & Environmental Science. 2010; **3**(6):700-714

[20] Acevedo S, Castillo J, Fernández A, Goncalves S, Ranaudo MA. A study of multilayer adsorption of asphaltenes on glass surfaces by photothermal surface deformation. Relation of this adsorption to aggregate formation in solution. Energy & Fuels. 1998;**12**(2):386-390

[21] Acevedo S et al. Relations between asphaltene structures and their physical and chemical properties: The rosary-type structure. Energy & Fuels. 2007;**21**(4): 2165-2175

[22] Adams JJ. Asphaltene adsorption, a literature review. Energy & Fuels. 2014; **28**(5):2831-2856

[23] Mullins OC. The asphaltenes. Annual Review of Analytical Chemistry. 2011;**4**: 393-418

[24] Mullins OC et al. Advances in asphaltene science and the Yen–Mullins model. Energy & Fuels. 2012;**26**(7): 3986-4003

[25] Mullins OC, Sheu EY, Hammami A, Marshall AG. Asphaltenes, Heavy Oils, and Petroleomics. Berlin, Germany: Springer Science & Business Media; 2007

[26] Nassar NN. Asphaltene adsorption onto alumina nanoparticles: Kinetics and thermodynamic studies. Energy & Fuels. 2010;**24**(8):4116-4122

[27] Gateau P, Hénaut I, Barré L, Argillier J. Heavy oil dilution. Oil & Gas Science and Technology. 2004;**59**(5): 503-509

[28] Ghanavati M, Shojaei M-J, Ahmad Ramazani SA. Effects of asphaltene content and temperature on viscosity of Iranian heavy crude oil: Experimental and modeling study. Energy & Fuels. 2013;**27**(12):7217-7232

[29] Ghannam MT, Hasan SW, Abu-Jdayil B, Esmail N. Rheological properties of heavy & light crude oil mixtures for improving flowability. Journal of Petroleum Science and Engineering. 2012;**81**:122-128

[30] Gomaa TMS, El-hoshoudy AN. Investigating the effect of different nanofluids types on crude oil viscosity. Petroleum & Petrochemical Engineering Journal. 2018;**2**(7):5

[31] Graue DJ. Upgrading and Recovery of Heavy Crude Oils and Natural Bitumens by In situ Hydrovisbreaking US Pat. No. 6,016,867. Texas: World Energy Systems Inc; 2001

[32] Guo J, Xiao H, Wu F. Prediction of viscosity of kerosene-based nanographene fluids by molecular dynamics simulation analysis. Journal of Molecular Liquids. 2023;**391**:123417

[33] Hasan SW, Ghannam MT, Esmail N. Heavy crude oil viscosity reduction and rheology for pipeline

transportation. Fuel. 2010;**89**(5): 1095-1100

[34] Franco C, Cardona L, Lopera S, Mejía J, Cortés F. Heavy oil upgrading and enhanced recovery in a continuous steam injection process assisted by nanoparticulated catalysts. In: SPE Improved Oil Recovery Conference. Tulsa, Oklahoma, USA: Society of Petroleum Engineers; 2016

[35] Franco CA, Zabala R, Cortés FB. Nanotechnology applied to the enhancement of oil and gas productivity and recovery of Colombian fields. Journal of Petroleum Science and Engineering. 2017;**157**:39-55

[36] Habibi A, Ahmadi M, Pourafshary P, Al-Wahaibi Y. Reduction of fines migration by nanofluids injection: An experimental study. SPE Journal. 2012; **18**(02):309-318

[37] Hosny R et al. Nanotechnology impact on chemical-enhanced oil recovery: A review and bibliometric analysis of recent developments. ACS Omega. 2023;**8**(49):46325-46345

[38] Hussein AK. Applications of nanotechnology in renewable energies— A comprehensive overview and understanding. Renewable and Sustainable Energy Reviews. 2015;**42**: 460-476

[39] Khalil M, Jan BM, Tong CW, Berawi MA. Advanced nanomaterials in oil and gas industry: Design, application and challenges. Applied Energy. 2017; **191**:287-310

[40] Taborda EA, Franco CA, Ruiz MA, Alvarado V, Cortés FB. Experimental and theoretical study of viscosity reduction in heavy crude oils by addition of nanoparticles. Energy & Fuels. 2017; **31**(2):1329-1338

[41] Nassar NN, Montoya T, Franco CA, Cortés FB, Pereira-Almao P. A new model for describing the adsorption of asphaltenes on porous media at a high pressure and temperature under flow conditions. Energy & Fuels. 2015;**29**(7): 4210-4221

[42] Franco CA, Montoya T, Nassar NN, Pereira-Almao P, Cortés FB. Adsorption and subsequent oxidation of colombian asphaltenes onto nickel and/or palladium oxide supported on fumed silica nanoparticles. Energy & Fuels. 2013; **27**(12):7336-7347

[43] Franco C, Patiño E, Benjumea P, Ruiz MA, Cortés FB. Kinetic and thermodynamic equilibrium of asphaltenes sorption onto nanoparticles of nickel oxide supported on nanoparticulated alumina. Fuel. 2013; **105**:408-414

[44] Tarboush BJA, Husein MM. Adsorption of asphaltenes from heavy oil onto in situ prepared NiO nanoparticles. Journal of Colloid and Interface Science. 2012;**378**(1):64-69

[45] Nassar NN, Hassan A, Pereira-Almao P. Effect of the particle size on asphaltene adsorption and catalytic oxidation onto alumina particles. Energy & Fuels. 2011;**25**(9):3961-3965

[46] Franco CA, Lozano MM, Acevedo S, Nassar NN, Cortés FB. Effects of resin I on asphaltene adsorption onto nanoparticles: A novel method for obtaining asphaltenes/resin isotherms. Energy & Fuels. 2015;**30**(1):264-272

[47] Lozano MM, Franco CA, Acevedo SA, Nassar NN, Cortés FB. Effects of resin I on the catalytic oxidation of n-C 7 asphaltenes in the presence of silica-based nanoparticles. RSC Advances. 2016;**6**(78):74630-74642

[48] Taborda Acevedo EA. Viscosity Reduction of Heavy Crude Oil through the Addition of Nanofluids on the Non-thermal Process. [Doctoral dissertation]. Medellín, Colombia: Universidad Nacional de Colombia sede Medellín; 2017

[49] Khan MR. Rheological properties of heavy oils and heavy oil emulsions. Energy Sources. 1996;**18**(4):385-391

[50] Moncayo-Riascos I, Taborda E, Hoyos BA, Franco CA, Cortés FB. Effect of resin/asphaltene ratio on the rheological behavior of asphaltene solutions in a de-asphalted oil and p-xylene: A theoretical–experimental approach. Journal of Molecular Liquids. 2020;**315**:113754

[51] Pinzón DM. Rheological demonstration of heavy oil viscosity reduction by NiO/SiO2 nanoparticles-assisted ultrasound cavitation. In: SPE Annual Technical Conference and Exhibition. Dallas, Texas, USA: Society of Petroleum Engineers; 2018

[52] Taborda EA, Alvarado V, Cortés FB. Effect of SiO2-based nanofluids in the reduction of naphtha consumption for heavy and extra-heavy oils transport: Economic impacts on the Colombian market. Energy Conversion and Management. 2017;**148**:30-42

[53] Taborda EA, Franco CA, Alvarado V, Cortés FB. A new model for describing the rheological behavior of heavy and extra heavy crude oils in the presence of nanoparticles. Energies. 2017;**10**(12): 2064

[54] Pal R, Rhodes E. A novel viscosity correlation for non-Newtonian concentrated emulsions. Journal of Colloid and Interface Science. 1985; **107**(2):301-307

[55] Huang X, Garcia MH. A Herschel–Bulkley model for mud flow down a slope. Journal of Fluid Mechanics. 1998; **374**:305-333

[56] Herschel WH. The change in viscosity of oils with the temperature. Industrial & Engineering Chemistry. 1922;**14**(8):715-722

[57] Caenn R, Darley HC, Gray GR. Composition and Properties of Drilling and Completion Fluids. Gulf Professional Publishing; 2011;**1**

[58] Hossain ME, Al-Majed AA. Fundamentals of Sustainable Drilling Engineering. New York, USA: John Wiley & Sons; 2015

[59] Amani M, Al-Jubouri M, Shadravan A. Comparative study of using oil-based mud versus water-based mud in HPHT fields. Advances in Petroleum Exploration Development. 2012;**4**(2):18-27

[60] Katende A et al. Improving the performance of oil based mud and water based mud in a high temperature hole using nanosilica nanoparticles. Colloids and Surfaces a: Physicochemical and Engineering Aspects. 2019;**577**:645-673

[61] Rafati R, Smith SR, Haddad AS, Novara R, Hamidi H. Effect of nanoparticles on the modifications of drilling fluids properties: A review of recent advances. Journal of Petroleum Science Engineering. 2018;**161**:61-76

[62] Adil A et al. Nanoparticle-based cutting fluids in drilling: a recent review. The International Journal of Advanced Manufacturing Technology. 2024;**131**(5): 2247-2264

[63] Villada Y, Iglesias MC, Casis N, Erdmann E, Peresin MS, Estenoz D.

Cellulose nanofibrils as a replacement for xanthan gum (XGD) in water based muds (WBMs) to be used in shale formations. Cellulose. 2018;**25**: 7091-7112

[64] Villada Y et al. Di-carboxylic acid cellulose nanofibril (DCA-CNF) as an additive in water-based drilling fluids (WBMs) applied to shale formations. Cellulose. 2021;**28**:417-436

[65] Bui B et al. Viscoelastic properties of oil-based drilling fluids. Annual Transactions of the Nordic Rheology Society. 2012;**20**:33-47

[66] Villada Y, Busatto C, Casis N, Estenoz D. Use of synthetic calcium carbonate particles as an additive in water-based drilling fluids. Colloids Surfaces A: Physicochemical Engineering Aspects. 2022;**652**:129801

[67] Fathima A, Almohsin A, Michael FM, Bataweel M, Alsharaeh EH. Polymer nanocomposites for water shutoff application-A review. Materials Research Express. 2018;**6**(3):032001

[68] Obino V, Yadav U. Application of polymer based nanocomposites for water shutoff—A review. Fuels. 2021;**2**(3): 304-322

[69] Tavakkoli O, Kamyab H, Shariati M, Mohamed AM, Junin R. Effect of nanoparticles on the performance of polymer/surfactant flooding for enhanced oil recovery: A review. Fuel. 2022;**312**:122867

[70] Al-Anssari S et al. Synergistic effect of nanoparticles and polymers on the rheological properties of injection fluids: Implications for enhanced oil recovery. Energy & Fuels. 2021;**35**(7):6125-6135

[71] Giraldo LJ et al. The effects of SiO_2 nanoparticles on the thermal stability and rheological behavior of hydrolyzed polyacrylamide based polymeric solutions. Journal of Petroleum Science and Engineering. 2017;**159**:841-852

[72] Zapata K, et al. Effect of Amine-Functionalized Nanoparticles (SiO2/ Amine) on HPAM Stability under Chemical Degradation Environments: An Experimental and Molecular Simulation Study. Energy & Fuels. 2023; **12**:8224-8236

[73] Guo P et al. Chemical water shutoff agents and their plugging mechanism for gas reservoirs: A review and prospects. Journal of Natural Gas Science Engineering. 2022;**104**:104658

[74] Thoniyot P, Tan MJ, Karim AA, Young DJ, Loh XJ. Nanoparticle–hydrogel composites: Concept, design, and applications of these promising, multi-functional materials. Advanced Science. 2015;**2**(1–2):1400010

[75] Chen L et al. Experimental investigation on the nanosilica-reinforcing polyacrylamide/ polyethylenimine hydrogel for water shutoff treatment. Energy & Fuels. 2018; **32**(6):6650-6656

[76] Shamlooh M, Hamza A, Hussein IA, Nasser MS, Salehi S. Reinforcement of polyacrylamide-co-tert-butyl acrylate base gel using nanosilica for conformance control at low and high reservoir temperatures. In: SPE International Conference and Exhibition on Formation Damage Control. Lafayette, Louisiana, USA: OnePetro; 2020

[77] Sheidaie A, Fahimpour J, Sharifi M. Experimental investigation of low-concentrated nanocomposite polymer gels for water shutoff treatment under reservoir conditions. SPE Journal. 2022; **27**(04):2390-2407

[78] Yudhowijoyo A, Rafati R, Sharifi Haddad A, Pokrajac D, Manzari M. Developing nanocomposite gels from biopolymers for leakage control in oil and gas wells. In: SPE Offshore Europe Conference and Exhibition. Aberdeen, UK: OnePetro; 2019

[79] Sydansk RD. Acrylamide-polymer/chromium (III)-carboxylate gels for near wellbore matrix treatments. SPE Advanced Technology Series. 1993; **1**(01):146-152

[80] Pérez-Robles S, Matute CA, Lara JR, Lopera SH, Cortés FB, Franco CA. Effect of nanoparticles with different chemical nature on the stability and rheology of acrylamide sodium acrylate copolymer/chromium (III) acetate gel for conformance control operations. Nanomaterials. 2019;**10**(1):74

[81] Pérez-Robles S, Cortés FB, Franco CA. Effect of the nanoparticles in the stability of hydrolyzed polyacrylamide/resorcinol/formaldehyde gel systems for water shut-off/conformance control applications. Journal of Applied Polymer Science. 2019;**136**(21):47568